I0014643

Robotics

The Impact of Robotics as a Service
(Original Steam Robots and Circuits to Design and Build)

Salvatore Harvell

Published By **Oliver Leish**

Salvatore Harvell

*Robotics: The Impact of Robotics as a Service
(Original Steam Robots and Circuits to Design and
Build)*

ISBN   978-1-998038-68-8

Legal & Disclaimer

Table Of Contents

Chapter 1: What is a Robot?

It's interesting to note that, despite all the buzz about robots, and all the advancements in robotics there's still no an established definition of what a robot is. It is true that there are the fundamental traits which a robot must be able to demonstrate and assist you in determining whether the object in question is robotic or non-robotic. The information will assist the way to decide what characteristics you should incorporate into your robot before you are able to say that it's a machine.

Four Basic Characteristics of a Robot

Robots have four fundamental features: movement, sense energy, intelligence and sensing.

Sensing

The basic idea is that a robot has to be able to sense the environment in a manner similar to human beings sense their environment. Robots may be able to sense their environment using lights that can mimic the eye's functions, or have chemical sensors which function as the nose, sonar sensors that resemble the nose, sensors for touch that mimic the skin, as well as sensory sensors for taste similar to tongue. These sensors assist the robot to become more aware of its surroundings and be able to recognize the changes in it.

Movement

A robot must be able to move by running on legs, getting on wheels or propellers. The robot has the ability to move, or even just a part of it like the arms, the head, or even just legs.

Intelligence

Robots should be outfitted with artificial intelligence, also known as AI. This usually happens through computer programming. This is why you will need an education in computer programming to equip your robot with understanding. It is essential to programme the robot's capabilities in order to make it aware of the things it can sense and when to be moved.

Energy

The robot must have the ability to power it. The source of energy can be chemical, electrical (battery) or solar. How the robot is energized is dependent on the task your robot needs to perform.

Working Definition of a Robot

In the interest of discussions as well as for reference purposes for reference, we define robots as machines which includes sensors, control systems manipulation equipment, software, and the power supply that is used to perform certain functions.

The construction of a robot is based on an knowledge of the basic principles in mechanical engineering math as well as physics and programming in computers. For certain situations there is also a need for specific expertise in biology, chemistry as well as medicine. While studying robotics, it is necessary to engage in a wide variety of disciplines in order to create robots that are able to solve particular issues.

A Brief History of Robotics

The term "robot" was first used in the play R.U.R (Rossum's Universal Robots) created in the year 1921 written by Czech playwright Karl Capek. The story is about a series of robots that were designed for use in factories but eventually rebel against the human bosses they were. The word robot is the Czech word that means slaves.

The word "robotics" came into use in the first novel. The Russian-born American writer Isaac Asimov used it in his piece "Runabout" (1942). In comparison with Capek, Asimov had a higher opinion about the place robots play in society. As a general rule, he identified robots as valuable machines which serve human beings and viewed them as an "better clean race. The three Laws of Robotics:

First Law of Robotics

Robots are not able to injure humans or, by lack of action, let a person to cause harm.

Second Law of Robotics

Robots must follow the instructions given by humans, except in cases where such commands would be contrary to the First Law.

Third Law of Robotics

Robots can protect themselves in the event that such safeguards does not infringe on laws of First Law or Second Law of Robotics.

Early Models of Robots

One of the first instances of a machine intended to complete a typical job was documented in the year 3000 BCE. Egyptian water clocks are paired by human statuettes that hit the hour bells to announce the passage of the clock. Around 400 BCE Archytus of Taremtum is credited for his invention of pulleys and screws, invented the first pigeon from wood which was adept at flying. Meanwhile, hydraulically-powered figurines that could speak prophecies were common during the Greek domination of Egypt during the second century BCE.

In the 1st century C.E., Petronius Arbiter constructed a doll capable of moving as humans. Giovanni Torriani, in 1557 constructed a wooden robotic, capable of bringing bread for the Emperor every day from the grocery store. In the 1700s, robots were commonplace, with a variety of impractical but incredibly sophisticated machines like steam-powered machines made in Canada and the well-known talking doll made developed by Thomas Edison. Although these inventions could be the inspiration for the style and function of robotics, the advancements made in the 20th century in the area of robotics was more advanced than previous innovations several times over.

The First Modern Robots

The robots we're familiar with were designed from George C. Devol in the 1950s. The inventor of Kentucky created and registered an reprogrammable device which he named "Unimate" derived from "Universal Automation." In the years that followed Devol tried to commercialize the product but was unsuccessful. But in 1960s, the entrepreneur-engineer Joseph Englberger bought the patent from Devol and modified it into an industrial robot. The company he founded was Unimation which was responsible for the production and selling of these items. The company was a success in his business, and as a matter of fact, Englberger is regarded today as the father of Robotics.

The field also advanced within educational institutions. Alan Turing, pioneering computer scientist, mathematician and philosopher, and cryptologist released his work "Computing Machiner and Intelligence" in which he suggested the Turing Test to test whether machines are able to be able to think on its own. The test is referred to by the name of Turing Test.

In the year 1958, Charles Rosen of the Stanford Research Institute created a research group to study the creation of a robot named "Shakey" that was more technologically advanced than Devol's Unimate. Shakey is able to move about the room, see the light with eyes "eyes" move around strange environments, in a certain degree respond to the changes happening to the environment around him. Shakey was referred to as Shakey due to his rapid and unsteady movements.

In the year 1966, at Massachusetts Institute of Technology (MIT), Joseph Weizenbaum created an artificial program called ELIZA that acts as a computer-based psychologist which manipulates the user's words to form queries.

In the year 1967, Richard Greenblatt developed MacHack the program equipped to play chess in reaction to an article published by Hubert Dreyfuss, in which Greenblatt claimed that no computer could defeat him in the game of chess. After the program was completed, Dreyfuss was invited to compete and was defeated. The program formed the basis of chess programming which later evolved to Big Blue, the program that beat grand master Gary Kasparov in 1997.

The fascination with robotics was one of the primary factors that led to the creation of computers. In 1964 it was the year that it was the IBM 360 becomes the first computer that was produced in a massive way.

Robots also play a key role for pioneering space exploration. In 1969 it was the year that NASA in the United States successfully used the modern technology of robotics and computers to assist Neil Armstrong in his lunar landing. Robots were also instrumental in the advancement of science. in 1994, Carnegie Universities crafted Dante II which is a walking robot that can successfully slide into the crater Mt Spur to gather samples of volcanic gas.

Commercial firms also capitalized on the appeal to mass audiences of robotics. As early as 1999 Sony launched its initial version AIBO the robotic dog which could be entertained, educated and converse with the owner. The AIBO has seen a number of improvements over the following years, and the most recent model, called the ERS-7M3, being released in 2005.

Honda has also launched the ASIMO robot which is an modern humanoid robot, in the year 2000. In 2004 Epsom was named the world's most compact robot (7 centimeters high) and weighing just 10 grams.) It is built for flying and recording footage during natural catastrophes.

Following its release in 2002, the robotic vacuum cleaner dubbed the Roomba was a massive success. The Roomba sold over 2.5 million units. This suggests that there's the demand for local robotic technology.

Chapter 2: Get Started

The initial step to build your robot is to decide what the robot is supposed to do which is the reason for making the machine. Robots are used in a variety of situations, but they are mostly designed to aid humans. It can aid you immensely to first understand the various functions and applications of robots.

The robots can be classified into two major groups: domestic and industrial robots.

Industrial Robots

Industrial robots are employed in factories for the production of goods with high-precision, like smartphones, automobiles, computers as well as medicines and food items. Robots improved productivity in various workplaces and led to the growth of industries. Each kind industrial robot comes with its distinct design and shape that best suits the purpose for which it is used. For example, robot arms typically are used in automobile assembly lines to spray paint on or join frames. They are the most commonly used robots of today. In recent times, robotics for agriculture were introduced for farm-related tasks, including cutting weeds, and harvesting crop.

Domestic Robots

Robots for domestic use are mostly used at home for household chores. They typically perform routine chores every day, such as cleaning floors, mowing the lawn, sweeping floors and many other jobs that individuals generally don't have the time complete. As an example, there are vacuum cleaners that are able to sweep floors. They're equipped with motion sensors to ensure that they do be able to avoid any obstacles. All you have to do is turn the switch and it will complete what it is supposed to do. It can collect dust and hairs from pets and serve for several hours.

Also, there are mower robots, which are able to trim lawns. They have sensors that detect edges of grass. Domestic robots can also be utilized for entertainment purposes, like Robosapien, AIBO, and iDog.

Choosing a Robotic Platform

The next stage in creating your robot is deciding on what type of robot you'd like to create. A typical robot design typically is based on "inspiration" of what the robot can do, and how it's going to be as.

The kinds of robots can be built are limitless. So long as you're able to imagine something that a robot could do, then you are able to build it. your goal. For beginners it is possible to start by learning these types of robots that include aquatic robots, stationary robots as well as hybrid robotics.

Land Robots

The land-based robots, especially ones that have wheels are among the most commonly motorized robots created by students due to the fact that they have a low investment, while giving the chance to understand more about robotics. However, the most sophisticated kind of robot is called the humanoid and is comparable to human beings. Humanoids need varying levels of freedom as well as coordination of motors from different manufacturers and utilize a variety of sensors.

Wheeled Robots

Wheels are one of the primary methods of adding mobility to the robot. They are utilized to move a variety of sizes of robots as well as robotic platforms. They can be of the same size and there's no limit on the number of wheels are available. Most often those robots with three wheels have two wheels with the caster is at one end. Advanced robots that have two wheels use the gyroscopic stabilizing system.

In addition, robots equipped with between four and six wheels typically have multiple motors to reduce the possibility of slippage. Mecanum wheels, also known as Omni-directional wheels may offer significant advantages to robots in mobility. The majority of people who are new to robotics do not realize that the cheapest DC motors will be able to mobilize robots of a medium dimensions. You will discover later There are a lot of other factors you should consider prior to adding motorization to your robotic.

Advantages

These robots can be used for novices as they're typically more affordable to build. They are simple in layout and build as well as a wide range of choices. Additionally, robotics that have at least six wheels could surpass the mobility of track systems.

Disadvantages

Wheeled robots generally have limited surface area of contact because only a tiny portion of the wheel's surface is touched by the ground. The result is a decrease in the traction, which can cause slippage.

Tracked Robots

Tracks are utilized in tanks to increase mobility. While tracks, commonly called treads don't offer the additional torque they reduce slippage, and also help distribute the robot's weight. This allows the robot to move around in loose soil like sand and gravel. Additionally tracks that are flexible can be able to navigate smoothly over bumpy terrain. Many hobbyists believe that tanks tracks look stylish compared to wheels.

Advantages

Solid contact with the floor prevents slippage that is common when using wheels. Tracks also spread the weight in a uniform manner, which aids the robot to navigate through diverse terrains. Tracks can be utilized to significantly increase the surface clearance of the machine without the need for a larger drive wheel.

Disadvantages

The biggest drawback of the use of a track system in robots is that when they turn it is possible to create damage to the surface, which can cause destruction to the tracks. Furthermore the robots usually are constructed around tracks as well as there's a limitation in the number of tracks available. Drive sprockets can additionally limit the amount of motors that can be used.

Legged Robots

Robots are increasingly employing legs to help them move. They are a good choice for robots that need to navigate on uneven ground. The majority of robot prototypes are designed with six legs which allows the robot to maintain its balance. The robots that have fewer legs are harder to balance since they require stable and dynamic balance. When the robot stops running at the midpoint of its stride, it is at risk of falling over. While there have been robotics moving with only one leg via hopping, bipeds Hexapods, and quadrupeds are the most popular types.

Advantages

The motion of the leg is most natural motion among platforms. It is able to quickly overcome large obstacles and traverse rough surfaces.

Disadvantages

Many beginners feel discouraged from creating their first robot which is able to move on legs because it is a complex task that requires mechanical, electronic and coding abilities. It is also necessary to locate the right battery size to give the necessary power which is why legless robots tend to be expensive to construct.

Aerial Robots

Humans are always influenced by the notion of flight. This extends to the realm of robotics. The concept that of Autonomous Unmanned Aerial Vehicle (AUAV) has grown in popularity through time, and numerous people have created numerous models. The benefits in creating these robots are not yet been able to overcome the drawbacks. While building aerial robots numerous hobbyists continue to use commercial remote control systems. Professional aircrafts like the Predator ordered from the US military, were partially autonomous however, recently updated version of Predator are able to complete aerial missions without human involvement.

Advantages

Aerial robotics are excellent for monitoring Remote controlled aircrafts have been built over time, and there is an extensive community of mechanics that can provide help and advice in building your personal aerial robots.

Disadvantages

The community is not as developed in the field of autonomy in control. The majority knowledge in this subject is governed from US military. US military. In addition, the type of robot can be expensive since the machine could break when you make a mistake in your calculations which could cause the machine to crash. an accident.

Aquatic Robots

Recently, increasing numbers groups, individuals, and businesses are developing unnamed water vehicles. However, there are many obstacles that need to be overcome to make the aquatic robot attractive to other robots communities. However, it's worth noting that there are businesses in the present that manufacture robots that clean the pools. The aquatic robots use ballasts, thrusters, wings as well as tails, fins, and wings for movement under water.

Advantages

A large portion of the ocean remains inaccessible, which means there's lots to explore if you decide to create aquatic robots that will aid in exploring the submerged world. Its design is sure to be distinctive and can have a trial run in the swimming pool.

Disadvantages

The aquatic robots can be expensive to make as well as there's the possibility that the machine might be lost in the sea. Also, it is worth noting that the majority of electronic components aren't compatible with sea water, specifically salty ones. It is also important to think about the pressure in water as moving over the deep sea will require a substantial expenditure and study. Also, there is a very small robotics community which can offer assistance and possibilities for wireless communications.

Hybrid Robots

Your robot's design might not fit in one of the categories listed previously or may consist from a number of different functional elements. It is important to note that this book was written to help the build of mobile robots, not ones which have fixed designs. When building hybrid designs It is recommended to go with a modular layout in which each component can be removed and tested separately.

Advantages

Hybrid robotics are developed and constructed based on the preferences of you and your requirements. They are able to be used for a range of purposes and may comprise modules. Hybrid robots can provide greater flexibility and capabilities.

Disadvantages

Hybrid robots can be difficult to make and cost a lot of money. The components must be tailored in order to suit the specific design.

Grippers and Arms

Although grippers and arms do not fall within the realm of robots that move, the field started with end-effectors and arms. Armes and grippers are best method for robots to communicate with the surroundings they are dealing with. The basic robotic arm could be equipped with only two or three movements and the more sophisticated ones may have more than dozen motions.

Chapter 3: Understanding Actuators

In the wake of learning about general facts about robots and robotics during the two previous chapters, it's time to select the appropriate actuators that will allow your robot to move.

What Are Actuators?

Actuators are the devices that convert energy into physical movement. For robotics, this energy typically comes from electricity. The majority of actuators in the present produce circular or linear motion. In this instance, for instance the term "DC motor" refers to a DC motor is one type of actuator.

The right choice of actuators to use with your robot is a matter of learning the different actuators that are available and an understanding of fundamental the mathematics and physical sciences.

Rotational Actuators

They convert electric energy into rotational motion. There are two fundamental mechanical variables that define every actuator (a) the speed of rotation which is usually measured in units of revolutions per minute, also known as rpm. (b) torque, or the force they are able to generate at a certain distance, which is usually expressed as the Oz*in, or N*m.

AC Motor

Alternating Current (AC) is not often utilized in robotics since the majority they are powered by Direct Current (DC) which is in the form of batteries or cells. Additionally, electronic components utilize DC which makes it more convenient to use the same kind of energy source for the actuators. AC motors are typically employed in industries that require high torque and where motors are attached to the wall outlet.

DC Motor

DC motors are typically cylindrical but are also available in various dimensions and shapes. Additionally, they have output shafts which rotate at the speed of 5000 to 10000 rpm. Although DC motors are extremely fast to rotate they are not all powerful. To reduce speed of rotation and boost the torque, a gear can be installed. In order to install a motor on an existing robot, you need to connect the motor's body to the frame of the robot. Therefore, most motors have mounting holes placed on the face of the motor to allow them to be simply put in. DC motors can turn counterclockwise or clockwise. The direction of rotation of the shaft that is turning could be observed using potentiometers or encoders.

Geared DC Motor

The DC Motor could be added using a gearbox in order to decrease the speed of the motor and increase its power. In the example above, if a DC motor is spinning at

5000 rpm, and generates the equivalent of 0.0005 N*m torque, using an gear ratio of 123:1 ("one hundred and twenty-three in one") gear reduces its speed by a ratio of 123 (resulting at 5000 rpm 40 rpm = 123) and also increase the torque by an amount of 123 (0.0005 per the 123 value equals 0.0615 N*m). The most popular varieties of gears are spur, planetary and the worm. Like an DC motor that is geared, a DC motor is also able to rotate in counter or clockwise direction. clockwise. It is possible to add an encoder onto the shaft, if you wish to track the number of turns the motor has made.

Hobby Servo Motors

Hobby Servo Motors, also called R/C Servo Motors are actuators that move to an angular location, and were previously utilized in larger remote controlled devices that controlled or steered flights surfaces. Nowadays, they're employed in various applications and the cost of them has been cut substantially, while the number of options is growing. The majority of servo motors are able to turn around 180 degrees. The hobby servo motor comprised of the DC motor, electronic the gears, as well as an instrument that determines the angle. It works in conjunction with electronics to move the motor to stop the shaft's output at the angle of a particular. They generally include three wires, current, control pulse and ground. The robot servo is a new servo, which offers the ability to provide feedback on position and also constant rotating. Servos are able to rotate clockwise as well as counterclockwise.

Stepper Motors

The name says it all that stepper motors turn in particular steps or degrees. The amount of degree that the shaft is rotated with each step may differ based upon a variety of factors. The majority of stepper motors do not contain gears. Therefore, as to the DC motor, torque can be described as to be low. The process of fixing gears to the stepper motor can have the same result as fixing gears to the DC motor.

Linear Actuators

Linear actuators produce linear movements. They possess three distinct mechanical characteristics: (a) the force that is measured in pounds or kilograms (b) speed expressed in milliseconds or inches/s. (c) the most and minimum distance rods can travel, which is and is also known as stroke, measured in millimeters or inches.

Linear DC Actuator

An linear DC actuator is typically made up of the DC motor that is connected by a screw that also rotates as the motor rotates. The lead screw is fitted with an incline that can be pushed to either the opposite direction or toward the motor, essentially changing the rotational motion into an essentially linear movement. A few DC linear actuators incorporate the linear potentiometer, which provides feedback for linear positions.

Solenoids

Solenoids comprise one coil wrapped around a central core. When the coil is powered up by the magnetic field, it pushes the core away from magnetic fields and produces a motion only in one direction. A number of coils or other mechanical arrangement will be required for the purpose of generating movements in various directions. The solenoid stroke can be extremely small, yet they are generally very swift. The strength is largely dependent on the dimensions of the coil as well as the power that flows through it.

Hydraulic and Pneumatic Actuators

Pneumatic and hydraulic actuators utilize either air or liquid to produce a linear movement. They can have long strokes, high-speed as well as high power. For these actuators to be used it is necessary to utilize an air or fluid compressor which makes them difficult to operate compared to standard electric actuators. These actuators are typically utilized in industrial processes due to their dimensions and speed of force.

Muscle Wire

Muscle wire is an specialized wire that contracts as the electricity flows through it. Once the power is out and the wire is cool, it returns to its length. The type of actuator described above isn't fast, powerful and produces a longer stroke. It is nevertheless among the best actuators to utilize when working using smaller components.

How to Choose the Proper Actuator for Your Robot

In order to help you choose the appropriate actuator for your specific task take a look at the following questions to guide you.

Note that technology and innovations are coming out, meaning there is no guarantee that something will last forever. Remember that an actuator can perform a variety of tasks within different contexts.

1. Are you in need of mobilizing an unwheeled robot?

Motors for drive should support the entire weight of the robot, and may require a gear reduction. A majority of robotics employ "skid steering" while trucks or cars use rack and pinion. If you're a fan of skid steering method, geared DC motors are suggested to utilize for machines with track or wheels. They provide continuous movement, as well as an adjustable position feedback via optical encoders. Since the required rotation is restricted to a specific angle, you could choose the hobby servo motor to perform stirring.

2. Do you have a maximum limit for the movement range?

If the angular range is restricted by 180 degrees, and the required torque isn't a major factor it is recommended to use a hobby servo. is suggested. The servo motor is offered in a variety of dimensions and torques, and come equipped with angular-position feedback. The majority of them use potentiometers. However, some special ones make use of optical encoders. These servos can be utilized to create smaller walking robots.

3. Do you require motors to move or turn large items?

Lifting a weight requires far greater power than the movement of a heavy object on the flat floor. It is more important to consider torque over speed. Consequently, it's best to choose an engineered gearbox that has a strong DC motor or linear DC actuator that has an excellent gear ratio. It is possible to use an actuator device that will stop the masses from crashing if the power source is interrupted. the source of power. It can be used with clamps or Worm gears.

4. Do you need your angle to be exact?

The stepper motors when paired with a motor controller can offer extremely precise and exact movement. They're more suitable when compared with servo motors due to their continuous rotation. There are, however, modern digital servo motors with high-end features with optical encoders that are capable of delivering high-precision.

5. Do you require the same movements with an unidirectional line?

Linear actuators are perfect to move parts, and positioning them along straight lines. They come in a variety of dimensions and configurations. If you want to move quickly such as pneumatics or solenoids. To achieve the highest torques, make use of linear DC actuators or hydraulics. If your movement needs a minimal amount of torque, it is possible to utilize muscles wire.

Chapter 4: Microcontrollers and Motor Controllers

Microcontrollers are regarded to be"the "brain" of the robot as they perform the entire process of decision-making, computations and communication. These are machines that have ability to run the program (a series of commands).

For interaction with the outside world, microcontrollers have several electrical signals (known by the name of pins) that can be turned off or on through programming functions. The pins also play a role to read electronic signals which are generated from sensors and other devices, and determining if they are high or low.

A majority of microcontrollers in the present can analyze analogue voltage signals or any signals that have the full spectrum of values instead of just two specific values by utilizing an analog digital converters or ADC. With the help of ADC an microcontroller is able to assign a numerical value an analog voltage which is not high nor low.

What Could Microcontrollers Do?

Many complex actions can be accomplished by setting pins at low and high and using them in an innovative way. But, creating complex algorithms like smart actions and data processing, or complex programs is not within the capabilities of microcontrollers because of their natural capacity and speed.

To light the blinker, you could create a sequence of repeating events in which microcontrollers raise a pin after which it will wait for several minutes, then turn it down and then wait for a few minutes, and then return to the beginning of the sequence. Lights that are attached to the pin will flash open-endedly.

Additionally, microcontrollers may be utilized to control of electronic devices, including actuators, when connected in motor controllers Bluetooth and WiFi interfaces as well as storage devices and several more. Due to its flexibility they can be found within everyday items. In essence, each electronic appliance or electrical gadget uses at least one microcontroller.

Similar to microprocessors within Central Processing Units in personal computer systems, microcontrollers aren't required accessories like external storage devices or RAM to run. So, even though microcontrollers have less power as microprocessors, creating circuits and devices that are based on microcontrollers is much simpler to do and much more cost-effective, since the hardware required is minimal. Microcontrollers produce a the smallest amount of electricity via pins. So, a general microcontroller can't power solenoids or drive electrical motors, big lights or other loads that are direct. In this way, it could result in physical harm for the microcontroller.

Programming Microcontrollers

There's no reason to shy at the programming of microcontrollers. It's not like the old days in which making a blinker demanded an in-depth knowledge of the microcontroller and at least a dozen lines of code, programming microcontrollers has become quite simple in the present. The easiest way to program a microcontroller is using the Integrated Development Environments (IDE) that utilizes modern programming language, complete line archives, which can cover every one of the actions that are most commonly performed along with a number of helpful examples that can help you start. Learn more about how to program your robot by reading Chapter 6.

How to Choose the Proper Microcontroller for Your Robot

It is necessary to have a microcontroller to control any robotic project, unless you're interested in BEAM robotics or wish to manage your robot using an R/C device or

Tether. In the beginning, picking the correct microcontroller may appear to be an overwhelming task especially when you consider the features, specifications as well as the uses. There are a variety of microcontrollers on the market these days, such as BasicATOM, POB Technology, Pololu, Arduino, BasicX and Parallax.

The following questions may help you to select the correct microcontroller to use:

1. What microcontrollers are widely utilized to control your robotics design?

Making robots isn't an issue of popularity However, the fact the microcontroller is part of a user base or has been utilized in a similar project makes the designing phase easy. By doing this, you'll be able to profit from experience of fellow enthusiasts. Hobbyists are often known to provide codes, photos as well as instructions and even learning lessons.

2. Are you in need of specific parts for a particular microcontroller?

If your robot is designed with special demands or needs particular component or accessory crucial to your project, choosing the right microcontroller for your needs is important. While most components can be linked to the majority of microcontrollers, certain accessories are specifically designed to connect with a specific microcontroller.

3. Are you looking for special functions to your robot?

The microcontroller must be capable of performing every one of the actions required for the robot to work efficiently. Certain features are shared by every microcontroller, for instance the ability to perform simple mathematical functions, using digital inputs and outputs and taking the right decisions. Other microcontrollers may require specific devices like PWM, ADC, and the ability to communicate with other protocols. Also, you should think about memory requirements, pin counts and speed specifications.

Motor Controllers

Motor controllers are electronic gadgets which serve as an intermediary device between the microcontroller, motors, as well as the power source.

While the microcontroller can decide the direction of travel and velocity of the motors it isn't able to drive them directly. The motor controller is able to deliver the necessary current in the right voltage however it doesn't have enough power to determine how fast the motor has to rotate.

Therefore, the microcontroller as well as the motor controller have to cooperate to ensure that the motors behave in accordance with. The microcontroller could provide instruction for the motor controller about how to charge motors using a standard and simple communication system like UART and PWM. Furthermore, some motor controllers can be manually controlled by an analog voltage, which is usually created by the use of a potentiometer.

The dimensions and weight of a motor control device can differ greatly from a controller smaller than end of a pencil, or a massive controller that weighs up to several kilograms. A controller's size and weight typically will have little impact to the robotic, except if you are building unknown aerial or underwater robots.

Chapter 5: Controlling Your Robot and Use of Sensors

According to our definition of a robot, it must be able to gather information about its environment as well as make intelligent decisions, and execute the actions based upon the results of its calculations. It also has the possibility that the robot can become semi-autonomous (with aspects that are under the control of humans as well as things it could do independently).

An excellent example is a sophisticated aquatic robot. The human is the one who controls the primary movements of the robot, while the processor installed measures and responds to underwater flow to ensure that the robot stays stationary while making sure that it doesn't drift. A camera that is installed inside the robot could send video to the person who controls it, while sensors can monitor water's temperature, pressure and much more. If communication is disrupted between the robotic and person, an auto-pilot program can take over and tell the robot to go to the area.

To control your robot you must determine your level of autonomy. The first step is to determine whether you would like the robot to be wireless, tethered or even autonomous.

Tethered

Direct Wired Control

The most straightforward method of controlling an automobile is with the handheld controller which is physically linked to the vehicle via a tether or cable. Switches, joysticks, knobs as well as levers and buttons within the controller let you control the robot, without having to install sophisticated electronic components. When this is the case it is possible that the power source and motors may be directly linked by a switch, allowing the speed of rotation. The machines are typically not equipped with artificial intelligence, and are seen as devices controlled by remotes rather instead of robots.

Wired Computer Control

Another option is to incorporate an electronic microcontroller in the computer while using the tether. Connecting the microcontroller through your computer's ports can allow users to control actions via a keyboard, keypad, a joystick or any other type of device. Incorporating a microcontroller into the robot's design may need programming to determine how your robot reacts to input.

Ethernet

Another option to utilize computer controlled robots is to connect the Ethernet interface. The robot directly connected to a router could be utilized for mobile robots. The creation of a robot that could communicate with the internet may be advanced, however typically, an internet connection wireless is recommended.

Wireless

Infrared

There is no need to throw away cords and wires when you are using infrared-based transmitters and receivers. It is usually a fantastic success for people who are new to. Infrared controls require "line of sight" to perform its function. The receiver must have the capability of seeing the transmitter in order to read the messages. Infrared remotes are a way to transmit commands to receivers infrared that come with microcontrollers which analyze these signals to control the movements by the machine.

Radio Frequency

Remote control devices typically use microcontrollers inside the receiver as well as transmitter to transmit data using radio frequency. The receiver box typically has an electronic circuit board (PCB) comprising the smallest motor controller servo and receiver. The RF signal requires an appropriate transmitter that is paired with either a transceiver or receiver. It doesn't require a clear line of sight. It can also offer a considerable distance. The simplest RF devices permit data transfers across devices at large distances. There's an unlimited distance that RF devices can provide.

Bluetooth

Bluetooth is a kind of Radio Frequency and follows certain protocols to send and receive information. It is a standard Bluetooth coverage is typically limited by 10 metres, but Bluetooth has the benefit of operating the robot through Bluetooth-enabled devices like smartphones, laptops, and PDAs. As with RF, Bluetooth provides two-way communications.

WiFi

New developments in wireless technology allows you control your robot using the Internet. For building an WiFi machine, it is necessary to need to be equipped with a wireless router which connects to the internet as well as the WiFi device that is installed on the robot. It is also possible to make use of a device that has been configured with TCP/IP and the help of a wireless router.

Autonomous

The high-level robots are fully self-contained. Thanks to recent advancements, it is now possible to use the microcontroller to its maximum capability and have it react to the inputs from sensors. Automated control comes in a variety of forms: limited sensors, programmed to provide zero feedback from the surrounding environment and more complex sensor feedback. True "autonomous" control includes different sensors as well as code that allows the robot to determine its own the most intelligent option to take at any time.

The most advanced ways of controlling that are currently being utilized by robots autonomous are visually and auditory commands. To control auditory the robot responds with the sounds of a human's voice to give instructions for example "get the ball" or "turn left." For visual commands, the robot could look at an object in order to determine what to take action. A robot's instructions to turn towards the left by showing an image of an arrow directed to the right is sophisticated programming. Although these tasks can be done but they do require advanced degree of programming, and typically thousands of hours.

Chapter 6: Assembling and Programming a Robot

Once you have learned about the essential elements of building your robot, the following phase is designing and creation of a frame to hold all of the parts together and gives the robot with a distinct appearance and form.

Constructing the Frame

There's no one-size-fits-all approach to making a frame because there's always a compromise that must be made. It is possible to choose lightweight frames, but it could be necessary to utilize expensive materials. There are many reasons to choose a robust or huge chassis, however you could find that it's expensive, laborious to build, or very heavy. It could be a complicated frame and could take a long amount of time to build and design.

Materials

There are various materials could be used to build the frame of your robot. If you experiment with different substances to build not just machines, but different types of machinery, you'll be able to recognize the benefits and drawbacks in what material is most appropriate for your particular project. The following list of suggested construction materials listed below is only the most common ones after you've experimented with a few of them then you are able to begin experimenting with mixing them together.

Basic Construction Materials

A few of the most basic material for construction could be used for frames of good quality. One of the cheapest is cardboard you will generally find on the cheap and can easily be cut, bent, layered or bent. It is possible to build a stronger cardboard box that appears more appealing and much more proportional to the dimensions of the robot. Then, you can paint it using epoxy or glue for a stronger structure, after which you can add an additional coats of paint.

Structural Flat Materials

To make a stronger frame it is possible to choose a common structural material for example, a sheet made of metal, plastic or even wood. It is just a matter of puncturing some holes for connecting the electronic parts. Wood that is stronger is likely to be thick and heavy and a thin piece of metal may become too pliable. There is the possibility of attaching components on each side while the wood can remain strong and solid.

If you're at a point in which you're ready get a frame from an outside source The best choice is to get the piece precisely cut using the use of lasers or water. Employing a third party to create the custom-designed part is advised only if you're certain of the measurements, as mistakes may cost you a lot. Businesses that provide computer-controlled cutting may also offer additional services like the painting process and bend.

3D Printing

The construction of a frame using 3D printed panels isn't necessarily the best structurally sound choice, due to the fact that it's constructed over multiple layers. But, it can create intricate and precise designs that are difficult to create using alternative processes. The 3D-printed part could include all of the necessary locations for the mounting of the electrical and mechanical parts and not affect the robot's weight. Over the last decade, the price of 3D printing was quite high however, as it gets more popular, the cost for manufacturing the parts is predicted to decrease.

Assembling the Parts Together

Based on the options available for the materials and techniques it is now time to begin making the pieces. The steps following to construct a straightforward design, appealing, and solid robot frame.

1. Select the materials you'd like to work with.

2. Collect all the components that your robot requires including electrical and mechanical components and then measure their dimensions. If you do not already have the necessary components then you may look up the dimensions which usually are supplied by manufacturer.

3. Create a variety of ideas to frame your picture. You aren't required to give specifics.

4. When you have found the right style, make sure that the structure is solid and that the components will be supported by the frame.

5. Draw every part of the robot using either paper or cardstock at the exact dimensions. Then, draw the elements in the CAD software and then print the parts.

6. Check the design using the computer and also in the actual settings using your prototype paper by testing the fit of every part and connecting.

7. Take another measurement until you're confident that your design is accurate, you can begin cutting your frame out of the substance. Note that you should measure 2 times. Only cut one time.

8. Make sure that every component is properly fitted prior to building the frame there is a need for modifications.

9. Build the frame with appropriate building materials, such as screws, glue, nails and duct tape or any other suitable binding tool you would like.

10. Put all the components in the frame, and voilà now you've created your very own robot!

Constructing the Robot Parts

The second step mentioned above must be explained further. In previous chapters, you've already selected the electrical components, which include the microcontroller, actuators and the motor controller. Next, you must build them in a way that they cooperate.

In this section, we'll be using common cable colors as well as terminal names that include the most common components. It is recommended to consult documents and datasheets when you're working on particular parts.

Attaching Motor Controllers to Motors

A geared DC motor, also known as linear DC actuator typically includes two wires, the red and black. Connect the black wire on the terminal M- of the DC motor controller, and connect the red wire on M+ terminal. If you connect the wires in the reverse direction will result in the motor to turn towards the reverse direction. For servo motors, there are three wires: black, red black and yellow. A motor controller for servos comes with pins which match the wires, so you could connect it straight away.

Attaching Microcontroller to Motor Controllers

Microcontrollers connect to motor controllers in several ways: 12C Serial, R/C or PWM. Make sure you consult the instruction manual of every microcontroller to find specific guidelines regarding the proper way to connect. Whatever choice you make the microcontroller as well as the motor controller's logic need to share the same GND reference. It is possible to achieve this through connecting the GND pins. Additionally, a level shifter will be required if these devices do not share the same levels of logic.

Attaching Batteries to a Microcontroller or a Motor Controller

Most motor controllers on the market today feature two screw terminals that are used for batteries, marked with B+ and -. If your batteries have an adapter and your controller has screw terminals, you can nevertheless look for an a-pair connector that has wires that can be connected onto screws terminals. If that's the case then you must find an alternative method to connect the battery with the motor controller, and you could still disconnect the battery and connect it to an charger. There is a chance that none of the mechanical and electrical components that you've chosen to run your robot can run on just one voltage. Therefore, they may require a variety of voltage regulators or batteries.

If you're building the robot using an Arduino microcontroller, DC gear motors and perhaps servo motors, it's obvious that batteries won't have the capacity to power each element completely. It is nevertheless recommended to select a battery that will directly power the gadgets as you require. The battery that has the highest capacity has to be connected to the motors that drive them. If, for instance, the drive motors you pick have an rated capacity of 12 volts then the primary battery should also have 12 Volts. This means you'll need the regulator to bring on the microcontroller with 5 volts. LiPo as well as NiMH batteries are among the best options for medium to small robots. Choose NiMH for those who require less expensive batteries, and LiPo when you require lightweight batteries. Make sure to remember that batteries are extremely powerful which can easily destroy your circuits if not connected properly. Be sure to ensure that the battery's polarity is right and your gadget can manage all the energy generated through the batteries. If you're not confident do not make any an assumption.

Adding Electrical Parts to Frame

Electrical components can be attached to your frame by a variety of techniques. It is important to ensure that the methods you employ, they don't transmit electricity. Common methods include screws Hex spacers Velcro double-sided tape glue, cable ties and many other.

Programming Your Robot

Programming is usually the final part of building your robot. If you've followed the instructions in preceding chapters, you've chosen the electronic parts like actuators motor controllers, microcontrollers sensors and much more. In this stage it is possible that you have created your own robots, and hopefully they look like what the one you would like it to. However, without a proper programming the robot will be nothing more than an awesome weight of paper.

There is a need for a second manual to learn about robotic programming. Instead, this chapter will help you understand how to begin and the things you need to know.

There are a variety of programming languages you could employ to programme the microcontroller which will act as your brain for the robot. Here are the top three commonly used programming languages that you could pick from:

Assembly

This language of programming is one step away from being fully-fledged computers, this makes it hard to master. It is a great language to utilize if you require complete instruction-level management of the robot.

Basic

Basic is among the most popular programming languages that robot enthusiasts use. The language is frequently used when developing microcontrollers mostly to create educational robots.

C++

C++ is a extremely widely used programming language. It is a top-level programming language, but it maintains a high lower-level control. One version to C++ is Processing with simplified programming for programming to simplify the process.

Java

Java is more advanced in comparison to C++ and offers security features that can overcome lower-level controls. A few manufacturers of microcontrollers like Parallax make components that are designed specifically for usage using Java.

Python

Python is among the most well-known languages used for scripting. It's simple to learn and can be utilized for quick and efficient integration of multiple applications.

If you've chosen the type of microcontroller that is geared towards hobbyists that is manufactured by a reputable company

There's a good chance there is a book to read through so you'll learn to program the controller in the programming language they prefer. If you prefer the microcontroller of a lesser producer, it's essential to determine which language the controller would like to utilize and which tools are offered.

Chapter 7: What Is A Robot?

In general, a robot can be defined as an electromechanical device which can react to its immediate environment in one way or the other, and autonomously take a decision or perform a task.

A robot's ability to take a decision and react to its environment is what distinguishes it from cars and toasters, for instance. These objects unlike a robot can't identify as they could not perceive or react autonomously to their environment.

The robot is commonly regulated by an electronic circuitry or computer program. A robotic platform or body is the part of a robot that determines how the robot will look and the role it will perform. The most commonly used robot platform is the wheeled type. It has some benefits when compared with other robot platforms.

Some of these benefits include things like comparatively low-cost and plenty of design options. They equally come with simple structure and design. The wheeled robotic platform is the best type of robot body for beginners.

Components Of A Robot

The brain of a robot is the microcontroller, it executes the program, makes the decision for the robot, makes computation and engages in communications. The Arduino UNO and Romeo are two most commonly microcontrollers used by DIY Arduino robots.

Robotic motors are devices that could convert electrical energy into mechanical energy. Motor drivers function as an intermediary devices between a microcontroller, a battery, and motors. It supplies the electric at a suitable voltage and helps the microcontroller to move the motors to move properly.

Robots' ability to perceive and react to their environment comes from their numerous types of electromechanical sensors. For instance, an infrared sensor could help an Arduino robot to identify how far it is from an abject. When it detects such distance, it feeds the information back into the microcontroller. A grayscale sensor, on the other hand, can be used to produce a line-tracking robot.

Robotics For Beginners: How To Build A Robot

In this part of the guide, we provide you with a step by step tip on how to make a simple robot. This is especially dedicated to DIY newbies who want to build their own Arduino robot. Having discussed the fundamental parts of a robot, this section will talk about the essential tools you need, how to assemble the robotic parts, program your robot and make it move.

So, let's begin.

What do you need to build a robot?

Tools Needed To Build A Robot

Screwdriver

You need a screwdriver for turning, driving in or eliminating screws and fasteners. It is better to purchase a screwdriver set that comes with different shapes and sizes, so you can handle different types of screws.

Soldering Pencil

You need a soldering pencil. It an essential part of any electrical work. You'd use it for soldering and de-soldering items on circuit boards. For this how to make robot tutorial, the soldering pencil will help us to solder the cables of the motor.

The tip of a soldering pencil is very hot when it is heated up. You must, therefore, be cautious about the way you handle it to prevent injury. If you're a first time user, take some time to get familiar with how to use the pencil before you begin working on your project.

Needle-Nose Pliers

Needle-nose pliers are freᵭuently utilized for cutting off the extra length of cables and wires. While pliers are commonly utilized for a project, they are not an essential part of this project.

Wire Stripper

A wire stripper is a hand-held device utilized for stripping the electrical insulation on wires. You can use the scissors as an alternative tool if you prefer. When utilizing a wire stripper, take care to only eliminate the first insulation layer. This exposes the lead wire inside and makes it easier for you to solder.

Assembling Your Robot: Step-By-Step Guide

Get an Arduino robot kit like Pirate: 4WD Arduino Mobile Robot Kit with Bluetooth 4.0 and follow the steps below to learn how to build a robot:

1.Assemble Your Robot Motor

Check the components bag for eight long screws. You will use these screws to hold your motors and keep them properly secure.

You will also find washers and gaskets in the robots component bag. Washers will help you to boost friction when fastening the motor and helps to keep it in place. The gaskets work for preventing the screw nuts from getting loose and falling when the robot is moving and when it collides with objects.

2.Solder the Cables

Look for black and red wires. They are included in the component bag. Fasten one black and one red wire with a length of 15 centimeters to each one of the four motors. After you have done this, use the wire stripper to remove the insulation from the two ends of the cables. Don't overdo it. When you're done, solder the wires on the pins attached to the motors. Repeat for the remaining four motors.

Mark the right positions for the red and black wires and solder accordingly.

3.Assemble the Romeo BLE Controller

Get three copper supports from the components bag. These supports are commonly one centimeter long. You will use them to secure the Romeo controller board. The controller board is made up of three holes. Put the three copper supports into those holes and fasten them into place with suitable screws.

4.Attach the battery to the Romeo BLE Controller

Get two countersunk screws with flat heads out of the component bag and attach the battery to the bottom of the car for your DIY robotics.

5.Design the Power Switch

Batteries are a significant part of robotic technology. The power switch helps you to limit or regulate your use of power. The power switch turns off power anytime the robot is not performing any task or in motion. So, put your power switch together and configure it. Be mindful of the gasket seꞅuence and screw nuts when putting your robots switch together.

When you have finished assembling the switch, begin to solder the wiring system of the switch. Use some of the wires you previously cut off. Strip the insulation from the two ends to expose the copper wire inside as you did in the previous step. Solder the stripped end of the wires to the switch pins. While doing this be conscious of the switch pins position. To do this:

•Link up a switch to the charger of the battery and note the exact position of these two items.

•Solder the red cables that connect the switch with the battery charger.

•to finish up, get a red cable and a black cable. Affix one end of the first wire to the negative pole of the battery charger and the other end of the second wire to the positive pole of the charger of the battery. Then connect the opposite ends of both wires to the Romeo BLE controller.

When you finish soldering it ensure you verify that the wiring system connecting the battery and Romeo controller is the same from beginning to end.

6.Design the Power Switch

With the use of eight M3 x 6mm screws, connect the side plates in the front and at the rear side of the bumper plates.

When fastening the screws don't fasten it completely in the beginning. This will make it easy for you to detach the upper board if you eventually need to make any alterations.

When you finish, re-connect the base plate to the body of your robotic car.

7.Connect the Motors With the Microcontroller Board

Link up your motors with the microcontroller board. Be careful when doing this. You should solder the red and black wires of the left motor into M2 and solder the red and black wires of the right motor into M1.

Your attention should be focused on the pack of the battery. You should solder the black wire into the GND wire port band solder the red wire in the wire port with the VND label. You need your screwdriver to lose or tight the wire ports. Ensure you fasten these ports well as soon as you insert the wires.

When you have finished soldering the motor wires into the microcontroller board, go ahead and connect the top plate to the bottom of the robotic car. You can fasten the sensor plate before connecting the top plate. Alternatively, you can skip this step and do it later on.

8.Connect your robot to an additional level

Look for the four holes on the bottom of the top plate. Screw the four M3 by 60mm Copper Standoffs in and connect an extra plate. You can utilize M3 x 6mm screws to connect the plate to the copper standoffs.

Add some wheels to the robotic platform you have developed and your robotic structure is set!

Chapter 8: Programming Your Robot

When you have finished building your robotic platform, the next step is to upload the microcontroller, so that your robot can move. Now, you have finished assembling your robot, it has all the essential features it needs to move.

What you need to do now is to check the sample codes for a file with the title- "MotorTest.ino", download it and upload the code to the microcontroller. Once you download and upload it, the motor will and wheels will start functioning almost immediately. If they don't work, verify if you install the batteries and power switch correctly.

As soon as you get the motors working, you have finish building your robot. You can now let your robotic car move. If the robotic car moves in a forward direction within seconds and moves backward in another second, you've got the component settings right. If not, you need to do some kinds of adjustments.

How to Make a Robot for Kids?

Robotics is nothing new to RootSaid, considering how well we are accustomed to using modern tech right from our childhood. Internet technology is nothing new to us, and the pandemic has even forced them to take our own education fully online. DIY Robots should've come naturally to our current generation, but it seems our education system is a bit slow on catching up with the latest trends in tech. So we usually find kids in their late teens getting their first real taste at robotics. Considering how widespread the usage of robots is in industries both big and small, their contribution to mass production and precise machining is unparalleled.

This enormous speed and efficiency that machines – especially robotic machines – bring in enhancing our lives have not been limited to industry floors or in high tech military applications. For example, we now see drones people use for recreation, which were once considered very difficult to build. The same goes for many other systems that were previously considered difficult to be made by enthusiasts with short budgets.

Just like any new technology, the DIY trend to tinker around and see what it offers has repeated for DIY Robot as well. This trend got accentuated by many factors: reduction in the prices for sensors, options for ꓱuick prototyping, availability of cheap microchips, and above all an ever-spreading supportive network of creators to help fellow tinkerers.

Another major contributor to the falling barrier of entry for DIY robotics enthusiasts is the entry of sensor-embedded garments, virtual reality modules, and similar high-tech devices that were previously unavailable for low prices. This trend is sure to continue in the coming years, with more and more advanced tech made available at approachable prices with the option for customization.

In India, the boom of internet technology in addition to computers and mobile phones paved the way for a wave of white-collar jobs that's still in high demand. We are now witnessing a new wave of cloud, AI, and robotics tech that's bringing in new opportunities in the hardware domain, also requiring appropriate programming skills to create turnkey solutions to novel challenges.

It is extremely important that we introduce these essential aptitudes to school-going children as well. Much of what the future of 21st century offers could perhaps only be tackled by adequate DIY knowledge, along with basic coding acumen for software and hardware projects. This judgment is based on research done by PeopleStrong, where it was found that one in four employment cuts globally would be from India. Thus, it is more than just good practice to introduce kids to robotics, it's high time.

Robotics for Beginners – What are the Parts of a Robot?

As this is a Robotics for Beginners Guide, I will try to make it as simple as possible. To make it more understandable, imagine a robot is like any other organism; like a human being.

We have sense organs. We see things with our eyes. We hear things with our ears, feel things with our skin, taste with our tongue, and smell with our nose. Our neurons transmit the signals from our sensors to our brain where we process them. Our brain process the signal takes a decision and sends the signal to our muscles to move our hands, legs or do whatever we want. Then we have a heart that circulates blood, supplies oxygen, provides energy for the working of your entire system. Then we have a body, where the sensors, muscles, heart, veins, and all other parts are neatly 'assembled'.

This is the same as in the case of a robot.

1.The Robot Chassis – The Body of the Robot

A robot has a chassis/ robot frame; which is similar to our body. A-frame that will support the whole robot. A system which is having enough space and is capable of handling the weight of all the sensors, power source, and all the cables used in the robot.

Qualities of a Good Robot Chassis for Kids

•Have enough space for all the components

•Provide good mechanical support for all the components

•Have mounting/screw holes where you can fix all the components in place

•Have enough strength to hold the entire weight of the components

•Here are some of the best robot chassis which are available online to make your own robot.

2.Sensors – The Sense Organs

Basically, sensors are specially designed devices or objects that will detect the properties, events, or changes in the environment, and then provide a corresponding signal. They are one of the crucial instruments which will bridge the physical and electronic world. There are different types of Sensors.

A robot has sensors; sensors such as IR sensors, which will sense infrared rays, ultrasonic sensors to sense ultrasonic waves, heat sensors to sense temperature, a pressure sensor to sense touch/pressure and so much more.

3.The Brain

So we now have all the data from the sensor. What to do with this sensor data? We need to process it. For that, we will have a processing unit. Normally, we will use a logic circuit.

This can be a simple circuit consisting of resistors, capacitors, transistors to perform simple logical decisions or a microcontroller such as an Arduino or PIC that can perform complex calculations. That depends upon the complexity of your robot.

Arduino and Raspberry Pi are the most commonly used controllers in the field of Robotics.

Arduino is simply an open-source base employing easy to use microcontroller boards with simple programming technique. Combining the power of strong microcontrollers and user-friendly IDEs which can be used to write and upload programs, which runs on your computers, less price, and less power consumption, it is effectively used in various industries, robotics, and home automation projects.

Raspberry Pi is a small credit card-sized, lightweight and compact computer which is used by both professions and hobbyist alike. This mini-computer can do (almost) anything a normal Linux machine can do. It is widely used in various industries, robotics, home automation projects.

There is a new version of Raspberry Pi – Raspberry Pi Pico which is similar to Arduino, available in market for Hobbyists. This board is very simple to begin with and start building your projects! If you are confused about whether you need to buy Raspberry Pi or Arduino for your next project, fear not. In this post – Raspberry Pi Pico or Arduino we have explained everything in detail so that you can choose which suits best for your project.

4.Muscles – The Actuators (Motors and other moving parts)

Once the signals are being processed, the result/reaction will be sent to the actuators. Actuators can be any electromechanical device such as motors that will move the motor, or servo motors that will lift the hand; or anything.

Motors are one of the most commonly used actuators used in robots. This is also the best actuator from where kids can start learning about the programming and moving their first robot.

Before going further with Robotics for Beginners, let us take a look at one of the most commonly used actuator – The Motor.

There are mainly 3 types of Motors

•DC Motor

•Servo Motor

•Stepper Motor

DC Motors are simple motors that have 2 terminals – one negative and one positive terminal. The speed of DC motors varies from the motor to motor and depends on the operating voltage and number of turns. One thing to notice here is, it is possible to change the direction of rotation of DC motor by reversing the polarity of the voltage provided at the terminals of the DC motor. For that, we can use an H Bridge circuit such as an L293D motor driver or other similar H Bridge circuit. They are widely used for driving wheels in a robot.

Servo Motors are compact and electronically controlled motors with +/-90 degree rotation. However, modifications can be made to get a complete 360-degree rotation. They are used in places where we need precise control over the movement such as robot hands, turrets, etc. There are 3 terminals to the servo motor.

•+ve – 5V

•-ve – Gnd

•Signal – PWM Signal

It is this PWM signal that determines the angle at which the servo motors should rotate.

Stepper Motors are basically DC motors that rotates in a step by step manner. Inside the stepper motor, the coils are wound in a specific pattern called phase. By energizing each coil seperately, we can precisely control the position and movement of the motor shaft. They are widely used in 3D printers.

5.Heart – The Battery

Now we almost have everything we need now. All we need is a power source. A power source that can provide enough energy to power up all the sensors, motors and the microcontroller. For that, we will use a current source such as a battery or a power adapter.

Chapter 9: Simple DIY Robots for Beginners

Now you know what a robot is and what are the parts of a DIY Robot. Now let's get practical, shall we? As you are all beginners in this field, in this session, I will show you and explain some of the basic, DIY Beginners robots which you can build and learn basic robotics. All of the equipment, tools and components used in these robots are cheap, simple and easily available at online markets. Give it a shot.

1.Line Follower Robot for Beginners

This is the easiest robot you can build for getting started with robotics. This robot uses IR or Infrared Sensor to 'See' a line in front of it and follow it automatically without human intervention. These line follower robots can be made with or without using microcontrollers.

How Line Follower Robot Works?

Simple Line follower Robot using IR sensors (mounted on either side of the robot) to sense the line, and uses these signals to drive DC motors using a DC Motor Driver. You can make a Beginners Line Follower with microcontroller or line follower without microcontroller.

What Will I Learn?

Here you will learn how IR sensor works, basic microcontroller coding and how to control DC motors using a motor driver IC.

2.Obstacle Avoidance Robot for Beginners

What is an Obstacle Avoiding Robot?

Obstacle avoidance robot is a simple robot equipped with UltraSonic Sensor which will automatically detect obstacles in front of it and change its direction by itself.

How an Obstacle Avoiding Robot Works?

An obstacle avoiding robot is able to sense whether some object is present in front of it using an ultrasonic transmitter/receiver pair. If there is any obstacle, the ultrasonic waves will hit the obstacle, bounce back and hit the receiver. This signal can be used to change the direction of the robot using DC motors.

What will I Learn?

You will learn to use the Basics of Robotics sensors like UltraSonic Sensors and Several motors such as DC motors and/or Servo Motors.

Remote Controlled Robot for Beginners

Remote controlled robots are those which can be controlled wirelessly using a remote controller. Here we will have a robot part and a remote controller part and the robot can be controlled using the remote controller.

How a Remote Controlled Robot Works?

Working of a Remote Controlled Robot is really simple. Here you will be having two parts – The Robot and the Remote controller. Here we will use a Wireless signal transmitter module in the Remote controller and a Receiver in the Robot. The signal from the remote controller is transmitted directly to the Robot using a Wireless module.

What will I learn?

You will learn the working of Wireless Communication modules such as Bluetooth, HC12, Zigbee etc and apply them to your Robot.

Robotics for Beginners – DIY Robot Example 1

As this is a Robotics for Beginners Guide, I will try to explain everything with an example. Consider this line follower robot using Arduino. This robot uses IR sensors to detect the track and uses motors to move around depending upon the color of the track. Here,

- IR sensor is the Sensor

- Arduino is the Brain

- Motors are the Muscles

- LiPo Battery is the Heart

- Robot Chassis the Body

Using all these parts, the robot senses the path, takes a decision on its own, and reacts to its environment by moving only through the black path.

So once again we are back with the same question. Now you know what all things we need to make make a robot for kids. But there are different types of sensors, motors, batteries, and microcontrollers available in the market. But all of them cannot be used to build a robot you want. For that, you will need a plan for a Robot.

Let us learn how to build a robot.

Step 1 – Plan Your Robot

First, decide what you want to build; a line follower, a pick and place robot or whatever you want. Decide what you want your robot to do.

For example – A Line follower that will follow a path without human intervention.

Step 2 – Know the Inputs

Understand what all are the inputs and outputs. For the robot to do whatever you want it to do, it should get maximum data from the environment, in order to process it. So find out what all things it need to know to perform that particular action.

In the case of our line follower, we need something that can sense the color of the path in front of it. For that, you can use an IR sensor or a color sensor.

Step 3 – Processing the Inputs and Decision Making

Once we have collected all the input parameters, we will have to feed it to the

Brain for processing. We have to process the inputs and make the robot to make the right decision in various conditions.

In our line follower, we have an Arduino microcontroller board which will act as its brain. The Arduino will read all the inputs from the IR sensors, process it and send the decisions/outputs to the motor driver board which will drive the motor.

Step 4 – The Output

Now the robot knows what to do. Now, all we need is to execute that action. For that, we will use an actuator. An actuator can be an electromechanical device.

In the case of our line follower, when the black line is curved towards left, it should go left.

When the black line turns right, the robot should go right. We have 2 DC motor which will drive the bot.

Step 5 – Power On the Robot

Our robot is almost ready. All you need is a power source to power everything up. Remember, you have to power all the devices including the sensors, microcontroller board, the motor drivers. the motors and all other actuators.

You have to choose the right power source for your robot. Different components have different voltage range for its working; below which it won't work or above which, it will burn off the component. So choosing the right power source can be a crucial task.

Step 6 – The Robot Chassis

Chassis is a frame where you mount all the parts of a robot together. You can either build one yourself or get one online.

10 Tips for Getting Started with Robotics

1.Learn about electronics

While this isn't one of the most fun parts about robotics, it is essential. For a while, I lived under the impression that I could do robotics without knowing anything about electronics. But, I found out that I was wrong pretty soon. Don't get me wrong, you don't have to have an EE degree, but you do need to know some of the basics. Getting Started In Electronics by Forrest Mimms is an excellent resource for this. You can find a review of this book here. There's also helpful online electronics tutorials.

2.Buy some books

In order to have a good start into robotics, you will need to start growing your library right off the bat. Getting the right books will provide invaluable help. Robot Building for Beginners is a good starting point. An absolute must-have book is Robot Builder's Bonanza. You'll also want to get some magazine subscriptions. Robot Magazine is great for beginners, along with Servo Magazine. You'll also find other interesting books, on our books page.

3.Start off small

This is probably one most important points of this whole article. Stay small! Resist the urge to let your mind run wild with possibilities of cooking robot that will dust and vacuum at the same time. You need to start off small. Try putting some motors onto a base (like some AOL CDs or a breadboard from Radio Shack or Jameco) and running them with a Basic Stamp or an OOPic. If you're more the kit type, you will find an impressive selection at RobotShop, Lynxmotion, Parallax, Rogue Robotics, and Budget Robotics. If you don't have any electronics or mechanics experience I'd recommend getting a kit.

4.Get LEGO Mindstorms if you don't have any programming experience

If you've never programmed before, you're in a bit of trouble, because you'll have to learn in order to do robotics, well, mostly. However, LEGO Mindstorms offers and an excellent resource for the totally illiterate. I have never heard anything bad about this product and HIGHLY recommend it. Plus, if you advance beyond its capabilities, there are tons of great websites and books about hacking it for other uses. You can buy the Mindstorms 2.0 kit here, or wait till Aug. 2006 to get the new version, Mindstorms NXT. VEX Robotics Kit is also a good starting point. I don't have any personal experience with it, but I've heard good things.

5.Enter a contest - I.E. Build a 'bot to do something

After you're initial robot or so, you'll need to start to plan for a robot that will actually do something. Part of the problem for a lot of people is that they never plan their robot ahead of time. When you have definite goals in mind, i.e. "I want my robot to patrol the house at night", you are much more motivated and interested in finishing. A great way to do this is to enter your robot into a contest. Mini Sumo, and the international Fire-Fighting Contest are excellent choices. Many clubs have annual contests and events.

6.Work regularly on your 'bots

Make yourself work on your robots regularly, especially if you're entering a contest! Coming back to a project after weeks of ignoring it is tough. Take that time to think about the project and plan. It will help, even if it's just for a few minutes before bed. Also, keep a regular journal of what you've done. Documenting your work is important.

7.Read about the mistakes of others

Take a look at our top mistakes with building a robot list and know what to avoid.

8.Don't be a tightwad

This is probably the second most important point in this article. Take it from one - Being a tightwad, or cheap person, isn't good. You may save a few dollars, but you will lose so much more with the extra time and frustration involved in being cheap. Don't get me wrong, you should always look for bargains, but if that involves desoldering components off of circuit boards, as opposed to spending $5 at Digi-Key, just give up. I've learned this lesson the hard way. Robotics isn't a cheap hobby, and sometime you'll have to face the facts. You're time and sanity are worth more.

9.Ask LOTS of questions

Sign up to our community and just ask questions. You'll learn more that way than from any book or website. Questions are never stupid. Don't be shy. No one ever gets good enough where they don't have to ask questions sometime. The forums at Robot Magazine are a good place to start. Also, the RS Community forum, here, especially the Let's Make Robot section here, which is dedicated to Project Ideas, Project Showcases, Beginners, etc.

10. Share your experiences with others

Don't make the rest of the world learn everything the hard way. That's the beauty of the internet. If you've figured something out, write an article, create a Tutorial or a Robot post. Let others know. Sheesh, that's the reason you're reading this right now, I'm letting you know how to do things the right way.

Robotics applications

Today, industrial robots, as well as many other types of robots, are used to perform repetitive tasks. They may take the form of a robotic arm, robotic exoskeleton or traditional humanoid robots.

Industrial robots and robot arms are used by manufacturers and warehouses, such as those owned by Amazon, Devol, Best Buy and more.

To function, a combination of computer programming and algorithms, a remotely controlled manipulator, actuators, control systems -- action, processing and perception

-- real-time sensors and an element of automation helps to inform what a robot or robotic system does.

Some additional applications for robotics are the following:

•home electronics -- see Honda's ASIMO

•computer science/computer programming

•artificial intelligence

•data science

•law enforcement/military

•mechanical engineering -- see Massachusetts Institute of Technology Robotics

•mechatronics

•nanotechnology

•bioengineering/healthcare

•aerospace -- see National Aeronautics and Space Administration's Urbie

Machine learning in robotics

Machine learning and robotics intersect in a field known as robot learning. Robot learning is the study of techniques that enable a robot to acquire new knowledge or skills through machine learning algorithms.

Some applications that have been explored by robot learning include grasping objects, object categorization and even linguistic interaction with a human peer. Learning can happen through self-exploration or via guidance from a human operator.

To learn, intelligent robots must accumulate facts through human input or sensors. Then, the robot's processing unit will compare the newly acquired data to previously stored information and predict the best course of action based on the data it has acquired.

However, it's important to understand that a robot can only solve problems that it is built to solve. It does not have general analytical abilities.

Chapter 10: Pre-Programmed Robots

Pre-programmed robots operate in a controlled environment where they do simple, monotonous tasks. An example of a pre-programmed robot would be a mechanical arm on an automotive assembly line. The arm serves one function — to weld a door on, to insert a certain part into the engine, etc. — and its job is to perform that task longer, faster and more efficiently than a human.

Humanoid Robots

Humanoid robots are robots that look like and/or mimic human behavior. These robots usually perform human-like activities (like running, jumping and carrying objects), and are sometimes designed to look like us, even having human faces and expressions. Two of the most prominent examples of humanoid robots are Hanson Robotics' Sophia (in the video above) and Boston Dynamics' Atlas.

Autonomous Robots

Autonomous robots operate independently of human operators. These robots are usually designed to carry out tasks in open environments that do not require human supervision. They are quite unique because they use sensors to perceive the world around them, and then employ decision-making structures (usually a computer) to take the optimal next step based on their data and mission. An example of an autonomous robot would be the Roomba vacuum cleaner, which uses sensors to roam freely throughout a home.

Examples Of Autonomous Robots

•Cleaning Bots (for example, Roomba)

•Lawn Trimming Bots

•Hospitality Bots

•Autonomous Drones

•Medical Assistant Bots

Teleoperated Robots

Teleoperated robots are semi-autonomous bots that use a wireless network to enable human control from a safe distance. These robots usually work in extreme geographical conditions, weather, circumstances, etc. Examples of teleoperated robots are the human-controlled submarines used to fix underwater pipe leaks during the BP oil spill or drones used to detect landmines on a battlefield.

Augmenting Robots

Augmenting robots either enhance current human capabilities or replace the capabilities a human may have lost. The field of robotics for human augmentation is a field where science fiction could become reality very soon, with bots that have the ability to redefine the definition of humanity by making humans faster and stronger. Some examples of current augmenting robots are robotic prosthetic limbs or exoskeletons used to lift hefty weights.

How do robots function?

Independent robots

Independent robots are capable of functioning completely autonomously and independent of human operator control. These typically require more intense programming but allow robots to take the place of humans when undertaking dangerous, mundane or otherwise impossible tasks, from bomb diffusion and deep-sea travel to factory automation. Independent robots have proven to be the most disruptive to society, eliminating low-wage jobs but presenting new possibilities for growth.

Dependent robots

Dependent robots are non-autonomous robots that interact with humans to enhance and supplement their already existing actions. This is a relatively new form of technology and is being constantly expanded into new applications, but one form of dependent robots that has been realized is advanced prosthetics that are controlled by the human mind.

A famous example of a dependent robot was created by Johns Hopkins APL in 2018 for a patient named Johnny Matheny, a man whose arm was amputated above the elbow. Matheny was fitted with a Modular Prosthetic Limb (MPL) so researchers could study its use over a sustained period. The MPL is controlled via electromyography, or signals sent from his amputated limb that controls the prosthesis. Over time, Matheny became more efficient in controlling the MPL and the signals sent from his amputated limb became smaller and less variable, leading to more accuracy in its movements and allowing Matheny to perform tasks as delicate as playing the piano.

Main components of a robot

Robots are built to present solutions to a variety of needs and fulfill several different purposes, and therefore, require a variety of specialized components to complete these tasks. However, there are several components that are central to every robot's construction, like a power source or a central processing unit. Generally speaking, robotics components fall into these five categories:

Control system

Computation includes all of the components that make up a robot's central processing unit, often referred to as its control system. Control systems are programmed to tell a robot how to utilize its specific components, similar in some ways to how the human brain sends signals throughout the body, in order to complete a specific task. These robotic tasks could comprise anything from minimally invasive surgery to assembly line packing.

Sensors

Sensors provide a robot with stimuli in the form of electrical signals that are processed by the controller and allow the robot to interact with the outside world. Common sensors found within robots include video cameras that function as eyes, photoresistors that react to light and microphones that operate like ears. These sensors allow the robot to capture its surroundings and process the most logical conclusion based on the current moment and allows the controller to relay commands to the additional components.

Actuators

As previously stated, a device can only be considered to be a robot if it has a movable frame or body. Actuators are the components that are responsible for this movement. These components are made up of motors that receive signals from the control system and move in tandem to carry out the movement necessary to complete the assigned task. Actuators can be made of a variety of materials, such as metal or elastic, and are commonly operated by use of compressed air (pneumatic actuators) or oil (hydraulic actuators,) but come in a variety of formats to best fulfill their specialized roles.

Power Supply

Like the human body requires food in order to function, robots require power. Stationary robots, such as those found in a factory, may run on AC power through a wall outlet but more commonly, robots operate via an internal battery. Most robots utilize lead-acid batteries for their safe qualities and long shelf life while others may utilize the more compact but also more expensive silver-cadmium variety. Safety, weight, replaceability and lifecycle are all important factors to consider when designing a robot's power supply.

Some potential power sources for future robotic development also include pneumatic power from compressed gasses, solar power, hydraulic power, flywheel energy storage organic garbage through anaerobic digestion and nuclear power.

End Effectors

End effectors are the physical, typically external components that allow robots to finish carrying out their tasks. Robots in factories often have interchangeable tools like paint sprayers and drills, surgical robots may be equipped with scalpels and other kinds of robots can be built with gripping claws or even hands for tasks like deliveries, packing, bomb diffusion and much more.

7 Advantages of Robots in the Workplace

Many people fear that robots or full automation may someday take their jobs, but this is simply not the case. Robots bring more advantages than disadvantages to the workplace. They enrich a company's ability to succeed while improving the lives of real, human employees who are still needed to keep operations running smoothly. If you're thinking about investing in some robots, share the advantages with your employees. You might be surprised at how many of them are quick to support the idea.

1.Safety

Safety is the most obvious advantage of utilizing robotics. Heavy machinery, machinery that runs at hot temperature, and sharp objects can easily injure a human being. By delegating dangerous tasks to a robot, you're more likely to look at a repair bill than a serious medical bill or a lawsuit. Employees who work dangerous jobs will be thankful that robots can remove some of the risks.

2.Speed

Robots don't get distracted or need to take breaks. They don't re🞎uest vacation time or ask to leave an hour early. A robot will never feel stressed out and start running slower. They also don't need to be invited to employee meetings or training session. Robots can work all the time, and this speeds up production. They keep your employees from having to overwork themselves to meet high pressure deadlines or seemingly impossible standards.

3.Consistency

Robots never need to divide their attention between a multitude of things. Their work is never contingent on the work of other people. They won't have unexpected emergencies, and they won't need to be relocated to complete a different time sensitive task. They're always there, and they're doing what they're supposed to do. Automation is typically far more reliable than human labor.

4.Perfection

Robots will always deliver quality. Since they're programmed for precise, repetitive motion, they're less likely to make mistakes. In some ways, robots are simultaneously an employee and a quality control system. A lack of quirks and preferences, combined with the eliminated possibility of human error, will create a predictably perfect product every time.

5.Happier Employees

Since robots are often assigned to perform tasks that people don't particularly enjoy, like menial work, repetitive motion, or dangerous jobs, your employees are more likely to be happy. They'll be focusing on more engaging work that's less likely to grind down their nerves. They might want to take advantage of additional educational opportunities, utilize your employee wellness program, or participate in an innovative workplace project. They'll be happy to let the robots do the work that leaves them feeling burned out.

6.Job Creation

Robots don't take jobs away. They merely change the jobs that exist. Robots need people for monitoring and supervision. The more robots we need, the more people we'll need to build those robots. By training your employees to work with robots, you're giving them a reason to stay motivated in their position with your company. They'll be there for the advancements and they'll have the unique opportunity to develop a new set of tech or engineering related skills.

7.Productivity

Robots can't do everything. Some jobs absolutely need to be completed by a human. If your human employees aren't caught up doing the things that could have easily be left for robots, they'll be available and productive. They can talk to customers, answer emails and social media comments, help with branding and marketing, and sell products. You'll be amazed at how much they can accomplish when the grunt work isn't weighing them down.

While we're still lightyears away from a fully robotic workplace, the robotic capabilities that many companies are currently utilizing have proven to be one of the greatest innovations of our time. Start by adding a few robots, and see where it takes you.

Chapter 5: The Disadvantages Of Robots

Onto the downsides of robots. Here are the robotics disadvantages that aren't quite as enjoyable.

1.They Lead Humans To Lose Their Job

Robots have a nasty habit of taking peoples' jobs.

I mean, in a capitalist system business owners have to do what it takes to maximize profits. And the brutal efficiency of robots makes them perfect for the task.

Humans just can't compete with a robot that can work 24/7 without making any mistakes. That fact can force people out of jobs they've done their entire lives.

2. They Need Constant Power

Robots need oodles of electricity to run.

That makes them expensive to run (more on this later) and potentially damaging to the environment.

Unless we shift over to greener sources of energy, the growing demand for robots in society could lead to additional issues with global warming and greenhouse gas emissions.

3. They're Restricted To Their Programming

Robots can't think for themselves (yet).

They rely on clever humans to program them for specific tasks. And, though artificial intelligence and machine learning are coming on fast, this is a limiting factor in what Robots can do.

They can't magic themselves up new capabilities at the drop of a hat. They can't think outside the box, mould themselves to novel needs, or do anything other than the task for which they're programmed.

4.The Perform Relatively Few Tasks

In a similar way, robots are only suited, as of now, for specific roles and responsibilities.

They come into their own in industry, research, medical practices, and the military. Outside of those domains, though, they have minimal practical usage.

Our day to day lives are slowly becoming more robot-centric. For now, though, there's a way to go before we start putting robots to work around the house at scale.

As realistic as they might become, a major disadvantage of robotics is their inability to feel, empathize, and interact as humans do.

It's another important factor on this list of pros and cons of robots.

As realistic as they might become, a major disadvantage of robotics is their inability to feel, empathize, and interact as humans do. It's another important factor on this list of pros and cons of robots.

5.They Have No Emotions

Robots don't feel anything either.

They're machines- clever chunks of metal with gears and gizmos keeping them 'alive'. But they could never feel any emotion that would allow them to empathise with, or relate to, what we're going through.

You could never sit down and have a heart to heart with a robot.

Sure, if the programming develops enough, then they might be able to say the right things, respond to particular cues, and react appropriately in a given situation.

But it wouldn't be real! They wouldn't actually be feeling anything beneath the surface.

6.They Impacts Human Interaction

Human interaction will suffer as robots become an increasing part of life.

Already, the rise of mobile phones has started this slippery slope. Just look around you in any public space and you see a mass of people staring at their screens.

We're more connected via the internet than ever before, but more isolated, lonely, and depressed too. We risk forgetting what it means to have an actual, real-life, human interaction.

There's every reason to believe that this is going to worsen as robots get an ever-greater role to play in daily life.

7.They Require Expertise To Set Them Up

Ask me to program a robot and I'd have no idea where to start.

Literally zilch.

In the future, it might become a common skill. But, for now, programming and setting up robots/computers for any given task remains something that requires particular expertise.

That puts the power into the hands of a limited number of skilled people around the world.

8.They're Expensive To Install And Run

Business owners looking to install robots in their factories/operations face significant upfront costs.

After all, robots aren't cheap- especially when they're high-tech, top of the line and needed for a specific task. The can put extreme financial pressure on an organisation.

Thanks to the electricity and manpower required to keep robots in working order, the running costs involved are high as well.

You have to hope that that the increases output justifies the initial investment.

9.They Mean Staff Need Retraining

The shift from manpower to robot-power doesn't always cull jobs for employees.

Sometimes it just requires a shift in responsibilities to operate the robots that have been brought in. For that to happen, though, employees will need to undergo significant retraining and upskilling.

10. They'll Cause Possible Class Divides

Expensive products/services are only available to people who can afford them.

Until robots become mass-produced, it's unlikely that everyday people will be able to buy them and reap the rewards they might offer. They'll be the reserve of the wealthiest people in society, who'll get another head start on the working class.

If robots can be used to extend peoples' wealth, then the divide between rich and poor will only increase.

11. They Suffer Expensive Faults And Repairs

Another financial issue involved with robots is to do with any faults and problems that occur.

Everything from the materials involved to the engineers required to fix them can be expensive. Throw in the cost of downtime and you're looking at significant sums of money.

That's bad news for business owners trying to recoup the cost of investment.

12. They Cause Cybersecurity Issues

Robots open the door to a range of cybersecurity problems too.

Even today, the rise of computers leaves many organisations and individuals open to attack. Hacks, ransomware, and identity theft are all potential hazards.

Now, fast forward a few decades to a time when robots are an everyday part of life.

They might be helping out around the house, caring for peoples' wellbeing, and running any number of key tasks. Imagine if somebody hacked their system and programmed them to 'misbehave'.

The potential repercussions could be extreme.

13. They Pose Potential Physical Danger From Malfunctions

It isn't uncommon for humans to work in close proximity with robots.

The last thing you want is for a malfunction to cause the robot to do something dangerous.

They're big heavy hunks of metal that pack immense power. It wouldn't take much for a broken robot to do serious damage to the lowly human stood next to it.

Of all potential robots disadvantages, one of the most likely is that they make humans overly reliant on them!

Of all potential robots disadvantages, one of the most likely is that they make humans overly reliant on them!

14. They Might Make Humans Overreliant On Robotic Help

Think about how much you use your phone.

There's a good chance it's the first thing you check in the morning and the last thing you see at night. Your smartphone may never leave your pocket or be far from your side.

You use it for everything from fact-finding and navigation to photography and contacting loved ones.

Imagine having to go a day without it! Most people would feel lost and unable to go about their day as normal.

We can expect the same thing to happen with robots.

As they become more widely-available, used, and accepted in society, we'll become ever more reliant on them. It'll be like the film Wall-e, where humans become overweight and entirely dependent on robots for support.

15. They May Reduce Human Capabilities

We've already started to take our phones for granted.

But imagine if the internet was suddenly turned off and we had to revert to a life without the World Wide Web. It would change absolutely everything!

When was the last time you wrote a letter? Or read an actual map? When was the last time you used the Yellow Pages? What about keeping track of your finances with a pen and paper? How about going to a restaurant without reading online reviews? When was the last time you went to the library to get a book?

The list of changes that would come about goes on and on and on.

Modern technology has revolutionised life. But it also detracts from other parts of it. We're less capable than we once were- less resilient, patient, and knowledgeable, to name just a few.

Imagine what would happen if we had robots doing everything for us! We'd forget what it meant to be human.

Chapter 11: What is Robotics ?

Robotics is that branch of engineering that deals with conception, design, operation, and manufacturing of robots. There was an author named Issac Asimov, he said that he was the first person to give robotics name in a short story composed in 1940's. In that story, Issac suggested three principles about how to guide these types of robotic machines. Later on, these three principles were given the name of Issac's three laws of Robotics. These three

laws state that:

Robots will never harm human beings.

Robots will follow instructions given by humans with breaking law one. Robots will protect themselves without breaking other rules.

There are many types of robots

they are used in many different environments and for many different uses.

Although being very diverse in application and form, they all share three basic similarities when it comes to their construction:

Robots all have some kind of mechanical

construction, a frame, form or shape designed to achieve a particular task. For example, a robot designed to travel across heavy dirt or mud, might use caterpillar tracks. The mechanical aspect is mostly the creator's solution to completing the assigned task and dealing with the physics of the environment around it. Form follows function. Robots have electrical components that power and control the machinery. For example, the robot with caterpillar tracks would need some kind of power to move the tracker treads. That power comes in the form of electricity, which will have to travel through a wire and originate from a battery, a basic electrical circuit. Even petrol powered machines that get their power mainly from petrol still require an electric current to start the combustion process which is why most petrol powered machines like cars, have batteries. The electrical aspect of robots is used for movement (through motors), sensing (where electrical signals are used to measure things like heat, sound, position, and energy status) and operation (robots need some level of electrical energy supplied to their motors and sensors in order to activate and perform basic operations) All robots contain some level of computer programming code. A program is how a robot decides when or how to do something. In the caterpillar track example, a robot that needs to move across a muddy road may have the correct mechanical construction and receive the correct amount of power from its battery, but would not go anywhere without a program telling it to move. Programs are the core essence of a robot, it could have excellent mechanical and electrical construction, but if its program is poorly constructed its performance will be very poor (or it may not perform at all). There are three different types of robotic programs: remote control, artificial intelligence and hybrid. A robot with remote control programming has a preexisting set of commands that it will only perform if and when it receives a signal from a control source, typically a human being with a remote control. It is perhaps more appropriate to view devices controlled primarily by human commands as falling in the discipline of automation rather than robotics. Robots that use artificial intelligence interact with their environment on their own without a control source, and can determine reactions to objects and problems they encounter using their preexisting programun ming. Hybrid is a form of programming that incorporates both AI and RC functions in them As more and more robots are designed for specific tasks this method of classification becomes more relevant. For example, many robots are designed for assembly work, which may not be readily adaptable for other applications. They are termed as "assembly robots". For seam welding, some suppliers provide complete welding systems with the robot i.e. the welding equipment along with other material handling facilities like turntables, etc. as an integrated unit. Such an integrated robotic system is called a "welding robot" even though its discrete manipulator unit could be adapted to a variety of tasks. Some robots are specifically designed for heavy load manipulation, and are labeled as "heavy-duty robots".

Current and potential applications include:

Military robots.

Industrial robots. Robots are increasingly used in manufacturing (since the 1960s). According to the Robotic Industries Association US data, in 2016 automotive industry was the main customer of industrial robots with 52% of total sales.[In the auto industry, they can amount for more than half of the "labor". There are even "lights off" factories such as an IBM keyboard manufacturing factory in Texas that was fully automated as early as 2003.

Cobots (collaborative robots).

Construction robots. Constructing

on robots can be separated into three types: traditional robots, robotic arm, and robotic exoskeleton.   Agricultural   robots (AgRobots).The use of robots in agriculture is closely linked to the concept of AI-assisted precision agriculture and drone usage.1996-1998 research also proved that robots can perform a herding task.

Medical robots of various types (such as da Vinci Surgical System and Hospi).

Kitchen automation. Commercial examples of kitchen automation are Flippy (burgers), Zume Pizza (pizza), Cafe X (coffee), Makr Shakr (cocktails), Frobot (frozen yogurts) and   Sally   (salads).Home   examples   are

Rotimatic (flatbreads baking) and Boris (dishwasher loading).

Robot combat for sport – hobby or sport event where two or more robots fight in an arena to disable each other. This has developed from a hobby in the 1990s to several TV series worldwide.

Cleanup of contaminated areas, such as toxic waste or nuclear facilities.

Domestic robots.

Nanorobots.

Swarm robotics.

Autonomous drones.

Sports field line marking.

Power source

At present, mostly (lead–acid) batteries are used as a power source. Many different types of batteries can be used as a power source for robots. They range from lead–

acid batteries, which are safe and have relatively long shelf lives but are rather heavy compared to silver–cadmium batteries that are much smaller in volume and are currently much more expensive. Designing a battery-powered robot needs to take into account factors such as safety, cycle lifetime and weight. Generators, often some type of internal combustion engine, can also be used. However, such designs are often mechanically complex and need a fuel, require heat dissipation and are relatively heavy. A tether connecting the robot to a power supply would remove the power supply from the robot entirely. This has the advantage of saving weight and space by moving all power generation and storage components elsewhere. However, this design does come with the drawback of constantly having a cable connected to the robot, which can be difficult to manage.Potential power sources could be:

pneumatic (compressed gases)

Solar power (using the sun's energy and converting it into electrical power)

hydraulics (liquids)

flywheel energy storage

organic garbage (through anaerobic digestion)

nuclear

Actuation

Actuators are the "muscles" of a robot, the parts which convert stored energy into movement.By far the most popular actuators are electric motors that rotate a wheel or gear, and linear actuators that control industrial robots in factories. There are some recent advances in alternative types of actuators, powered by electricity, chemicals, or compressed air.

Electric motors

The vast majority of robots use electric motors, often brushed and brushless DC motors in portable robots or AC motors in industrial robots and CNC machines. These motors are often preferred in systems with lighter loads, and where the predominant form of motion is rotational.

Linear actuators

Various types of linear actuators move in and out instead of by spinning, and often have quicker direction changes, particularly when very large forces are needed such as with industrial robotics. They are typically powered by compressed and oxidized air (pneumatic actuator) or an oil (hydraulic actuator) Linear actuators can also be powered by electricity which usually consists of a motor and a leadscrew. Another common type is a mechanical linear actuator that is turned by hand, such as a rack and pinion on a car.

Chapter 12: Series elastic actuators

Series elastic actuation (SEA) relies on the idea of introducing intentional elasticity between the motor actuator and the load for robust force control. Due to the resultant lower reflected inertia, series elastic actuation improves safety when a robot interacts with the environment (e.g., humans or workpiece) or during collisions. Furthermore, it also provides energy efficiency and shock absorption (mechanical filtering) while reducing excessive wear on the transmission and other mechanical components. This approach has successfully been employed in various robots, particularly advanced manufacturing robots and walking humanoid robots. he controller design of a series elastic actuator is most often performed within the passivity framework as it ensures the safety of interaction with unstructured environments. Despite its remarkable stability robustness, this framework suffers from the stringent limitations imposed on the controller which may trade-off performance.The reader is referred to the following survey which summarizes the common controller architectures for SEA along with the corresponding sufficient passivity conditions.One recent study has derived the necessary and sufficient passivity conditions for one of the most common impedance control architectures, namely velocity-sourced SEA.This work is of particular importance as it drives the non-conservative passivity bounds in an SEA scheme for the first time which allows a larger selection of control gains.

Air muscles

Pneumatic artificial muscles, also known as air muscles, are special tubes that expand(typically up to 40%) when air is forced inside them. They are used in some robot applications.

Muscle wire

Muscle wire, also known as shape memory alloy, Nitinol® or Flexinol® wire, is a material which contracts (under 5%) when electricity is applied. They have been used for some small robot applications.

Electroactive polymers

EAPs or EPAMs are a plastic material that can contract substantially (up to 380% activation strain) from electricity, and have been used in facial muscles and arms of humanoid robots,and to enable new robots to float, fly, swim or walk.

Piezo motors

Recent alternatives to DC motors are piezo motors or ultrasonic motors. These work on a fundamentally different principle, whereby tiny piezoceramic elements, vibrating many thousands of times per second, cause linear or rotary motion. There are different mechanisms of operation; one type uses the vibration of the piezo elements to step the motor in a circle or a straight line. Another type uses the piezo elements to cause a nut to vibrate or to drive a screw. The advantages of these motors are nanometer resolution, speed, and available force for thl Elastic nanotubes are a promising artificial muscle technology in early-stage experimental development. The absence of defects in carbon nanotubes enables these filaments to deform elastically by several percent, with energy storage levels of perhaps 10 J/cm3 for metal nanotubes. Human biceps could be replaced with an 8 mm diameter wire of this material. Such compact "muscle" might allow future robots to outrun and outjump humans.

Sensing

Sensors allow robots to receive information about a certain measurement of the environment, or internal components. This is essential for robots to perform their tasks, and act upon any changes in the environment to calculate the appropriate response. They are used for various forms of measurements, to give the robots warnings about safety or malfunctions, and to provide real-time information of the task it is performing.

Touch

Current robotic and prosthetic hands receive far less tactile information than the human hand. Recent research has developed a tactile sensor array that mimics the mechanical properties and touch receptors of human fingertips.The sensor array is constructed as a rigid core surrounded by conductive fluid contained by an elastomeric skin. Electrodes are mounted on the surface of the rigid core and are connected to an impedance-measuring device within the core. When the artificial skin touches an object the fluid path around the electrodes is deformed, producing impedance changes that map the forces received from the object. The researchers expect that an important function of such artificial

fingertips will be adjusting robotic grip on held objects.

Scientists from several European countries and Israel developed a prosthetic hand in 2009, called SmartHand, which functions like a real one allowing patients to write with it, type on a keyboard, play piano and perform other fine movements. The prosthesis has sensors which enable the patient to sense real feeling in its fingertips.

Vision

Computer vision is the science and technology of machines that see. As a scientific discipline, computer vision is concerned with the theory behind artificial systems that extract information from images. The image data can take many forms, such as video sequences and views from cameras. In most practical computer vision applications, the computers are pre-programmed to solve a particular task, but methods based on learning are now becoming increasingly common. Computer vision systems rely on image sensors which detect electromagnetic radiation which is typically in the form of either visible light or infra-red light. The sensors are designed using solid-state physics. The process by which light propagates and reflects off surfaces is explained using optics. Sophisticated image sensors even require quantum mechanics to provide a complete understanding of the image formation process. Robots can also be equipped with multiple vision sensors to be better able to compute the sense of depth in the environment. Like human eyes, robots' "eyes" must also be able to focus on a particular area of interest, and also adjust to variations in light intensities. There is a subfield within computer vision where artificial systems are designed to mimic the processing and behavior of biological system, at different levels of complexity.

Also, some of the learning-based methods developed within computer vision have their background in biology.

Other

Other common forms of sensing in robotics use lidar, radar, and sonar. MkLidar measures distance to a target by illuminating the target with laser light and measuring the reflected light with a sensor. Radar uses radio waves to determine the range, angle, or velocity of objects. Sonar uses sound propagation to navigate, communicate with or detect objects on or under the surface of the water.

Chapter 13: Manipulation

A definition of robotic manipulation has been provided by Matt Mason as: "manipulation refers to an agent's control of its environment through selective contact". Robots need to manipulate objects; pick up, modify, destroy, or otherwise have an effect. Thus the functional end of a robot arm intended to make the effect (whether a hand, or tool) are often referred to as end effectors,(while the "arm" is referred to as a manipulator).Most robot arms have replaceable end-effectors, each allowing them to perform some small range of tasks. Some have a fixed manipulator that cannot be replaced, while a few have one very general purpose manipulator, for example, a humanoid hand.

Mechanical grippers

One of the most common types of end-effectors are "grippers". In its simplest manifestation, it consists of just two fingers that can open and close to pick up and let go of a range of small objects. Fingers can, for example, be made of a chain with a metal wire run through it.Hands that resemble and work more like a human hand include the Shadow Hand and the Robonaut hand.Hands that are of a mid-level complexity include the Delft hand.Mechanical grippers can come in various types, including friction and encompassing jaws. Friction jaws use all the force of the gripper to hold the object in place using friction. Encompassing jaws cradle the object in place, using less friction.

Suction end-effectors

Suction end-effectors, powered by vacuum generators, are very simple astrictive devices that can hold very large loads provided the prehension surface is smooth enough to ensure suction.

Pick and place robots for electronic components and for large objects like car windscreens, often use very simple vacuum end-effectors. Suction is a highly used type of end-effector in industry, in part because the natural compliance of soft suction end-effectors can enable a robot to be more robust in the presence of imperfect robotic perception. As an example: consider the case of a robot vision system estimates the position of a water bottle, but has 1 centimeter of error. While this may cause a rigid mechanical gripper to puncture the water bottle, the soft suction end-effector may just bend slightly and conform to the shape of the water bottle surface.

General purpose effectors

Some advanced robots are beginning to use fully humanoid hands, like the Shadow Hand, MANUS,and the Schunk hand.These are highly dexterous manipulators, with as many as 20 degrees of freedom and hundreds of tactile sensors.

Locomotion

Rolling robots

simplicity, most mobile robots have four wheels or a number of continuous tracks. Some researchers have tried to create more complex wheeled robots with only one or two wheels. These can have certain advantages such as greater efficiency and reduced parts, as well as allowing a robot to navigate in confined places that a four-wheeled robot would not be able to.

Two-wheeled balancing robots

Balancing robots generally use a gyroscope to detect how much a robot is falling and then drive the wheels proportionally in the same direction, to counterbalance the fall at hundreds of times per second, based on the dynamics of an inverted pendulum.Many different balancing robots have been designed. While the Segway is not commonly thought of as a robot, it can be thought of as a component of a robot, when used as such Two robot snakes. Left one has 64 motors (with 2 degrees of freedom per segment), the right one 10. Several snake robots have been successfully developed. Mimicking the way real snakes move, these robots can navigate very confined spaces, meaning they may one day be used to search for people trapped in collapsed buildings.The Japanese ACM-R5 snake robots can even navigate both on land and in water.

Robotic Fish: iSplash-II

In 2014 iSplash-II was developed by PhD student Richard James Clapham and Prof. Huosheng Hu at Essex University. It was the first robotic fish capable of outperforming real carangiform fish in terms of average maximum velocity (measured in body lengths/ second) and endurance, the duration that top speed is maintained. This build attained swimming speeds of 11.6BL/s (i.e. 3.7 m/s).The first build, iSplash-I (2014) was the first robotic platform to apply a full-body length carangiform swimming motion which was found to increase swimming speed by 27% over the traditional Radar,

GPS, and lidar, are all combined to provide proper navigation and obstacle avoidance (vehicle developed for 2007 DARPA Urban Challenge) approach of a posterior confined waveform.

Sailing

The autonomous sailboat robot Vaimos Sailboat robots have also been developed in order to make measurements at the surface of the ocean. A typical sailboat robot is Vaimosbuilt by IFREMER and ENSTA-Bretagne. Since the propulsion of sailboat robots uses the wind, the energy of the batteries is only used for the computer, for the communication and for the actuators (to tune the rudder and the sail). If the robot is equipped with solar panels, the robot could theoretically navigate forever. The two main competitions of sailboat robots are WRSC, which takes place every year in Europe, and Sailbot.

Environmental interaction and navigation

of skating robots have been developed, one of which is a multi-mode walking and skating device. It has four legs, with unpowered wheels, which can either step or roll.Another robot, Plen, can use a miniature skateboard or roller-skates, and skate across a desktop.

Chapter 14: Capuchin, a climbing robot
Climbing

Several different approaches have been used to develop robots that have the ability to climb vertical surfaces. One approach mimics the movements of a human climber on a wall with protrusions; adjusting the center of mass and moving each limb in turn to gain leverage. An example of this is Capuchin,built by Dr. Ruixiang Zhang at Stanford University, California. Another approach uses the specialized toe pad method of wall-climbing geckoes, which can run on smooth surfaces such as vertical glass. Examples of this approach include Wallbot and Stickybot. China's Technology Daily reported on 15 November 2008, that Dr. Li Hiu Yeung and his research group of New Concept Aircraft (Zhuhai) Co., Ltd. had successfully developed a bionic gecko robot named "Speedy Freelander". According to Dr. Yeung, the gecko robot could rapidly climb up and down a variety of building walls, navigate through ground and wall fissures, and walk upside-down on the ceiling. It was also able to adapt to the surfaces of smooth glass, rough, sticky or dusty walls as well as various types of metallic materials. It could also identify and circumvent obstacles automatically. Its flexibility and speed were comparable to a natural gecko. A third approach is to mimic the motion of a snake climbing a pole.

Swimming (Piscine)

It is calculated that when swimming some fish can achieve a propulsive efficiency greater than 90%.Furthermore, they can accelerate and maneuver far better than any man-made boat or submarine, and produce less noise and water disturbance. Therefore, many researchers studying underwater robots would like to copy this type of locomotion. Notable examples are the Essex University Computer Science Robotic Fish G9, and the Robot Tuna built by the Institute of Field Robotics, to analyze and mathematically model thunniform motion. The Aqua Penguin, designed and built by Festo of Germany, copies the streamlined shape and propulsion by front "flippers" of penguins. Festo have also built the Aqua Ray and Aqua Jelly, which emulate the locomotion of manta ray, and jellyfish, respectively. Segway refer to them as RMP (Robotic Mobility Platform). An example of this use has been as NASA's Robonaut that has been mounted on a Segway.

One-wheeled balancing robots

A one-wheeled balancing robot is an extension of a two-wheeled balancing robot so that it can move in any 2D direction using a round ball as its only wheel. Several one-wheeled balancing robots have been designed recently, such as Carnegie Mellon University's "Ballbot" that is the approximate height and width of a person, and Tohoku Gakuin University's "BallIP".Because of the long, thin shape and ability to maneuver in tight spaces, they have the potential to function better than other robots in environments with people.

Spherical orb robots

Several attempts have been made in robots that are completely inside a spherical ball, either by spinning a weight inside the ball,or by rotating the outer shells of the sphere.These have also been referred to as an orb botor a ball bot.

Six-wheeled robots

Using six wheels instead of four wheels can give better traction or grip in outdoor terrain such as on rocky dirt or grass.

Tracked robots

TALON military robots used by the United States Army Tank tracks provide even more traction than a six-wheeled robot. Tracked wheels behave as if they were made of hundreds of wheels, therefore are very common for outdoor and military robots, where the robot must drive on very rough terrain. However, they are difficult to use indoors such as on carpets and smooth floors. Examples include NASA's Urban Robot "Urbie".

Walking applied to robots

Walking is a difficult and dynamic problem to solve. Several robots have been made which can walk reliably on two legs, however, none have yet been made which are as robust as a human. There has been much study on human inspired walking, such as AMBER lab which was established in 2008 by the Mechanical Engineering Department at Texas A&M University.Many other robots have been built that walk on more than two legs, due to these robots being significantly easier to construct.Walking robots can be used for uneven terrains, which would provide better mobility and energy efficiency than other locomotion methods. Typically, robots on two legs can walk well on flat floors and can occasionally walk up stairs. None can walk over rocky, uneven terrain. Some of the methods which have been tried are:

ZMP technique

The zero moment point (ZMP) is the algorithm used by robots such as Honda's ASIMO. The robot's onboard computer tries to keep the total inertial forces (the combination of Earth's gravity and the acceleration and deceleration of walking), exactly opposed by the floor reaction force (the force of the floor pushing back on the robot's foot). In this way, the two forces cancel out, leaving no moment (force causing the robot to rotate and fall over).However, this is not exactly how a human walks, and the difference is obvious to human observers, some of whom have pointed out that ASIMO walks as if it needs the lavatory.ASIMO's walking algorithm is not static, and some dynamic balancing is used (see below). However, it still requires a smooth surface to walk on.

Hopping

Several robots, built in the 1980s by Marc Raibert at the MIT Leg Laboratory, successfully demonstrated very dynamic walking. Initially, a robot with only one leg, and a very small foot could stay upright simply by hopping. The movement is the same as that of a person on a pogo stick. As the robot falls to one side, it would jump slightly in that direction, in order to catch itself.Soon, the algorithm was generalised to two and four legs. A bipedal robot was demonstrated running and even performing somersaults. A quadruped was also demonstrated which could trot, run, pace, and bound.For a full list of these robots, see the MIT Leg Lab Robots page.

Chapter 15: Dynamic balancing (controlled falling)

A more advanced way for a robot to walk is by using a dynamic balancing algorithm, which is potentially more robust than the Zero Moment Point technique, as it constantly monitors the robot's motion, and places the feet in order to maintain stability. This technique was recently demonstrated by Anybots' Dexter Robot,which is so stable, it can even jump.Another example is the TU Delft Flame.

Passive dynamics

Perhaps the most promising approach utilizes passive dynamics where the momentum of swinging limbs is used for greater efficiency. It has been shown that totally unpowered humanoid mechanisms can walk down a gentle slope,using only gravity to propel themselves. Using this technique, a robot need only supply a small amount of motor power to walk along a flat surface or a little more to walk up a hill. This technique promises to make walking robots at least ten times more efficient than ZMP walkers, like ASIMO.

Other methods of locomotion

Flying

A modern passenger airliner is essentially a flying robot, with two humans to manage it. The autopilot can control the plane for each stage of the journey, including takeoff, normal flight, and even landing. Other flying robots are uninhabited and are known as unmanned aerial vehicles (UAVs). They can be smaller and lighter without a human pilot on board, and fly into dangerous territory for military surveillance missions. Some can even fire on targets under command. UAVs are also being developed which can fire on targets automatically, without the need for a command from a human. Other flying robots include cruise missiles, the Entomopter, and the Epson micro helicopter robot. Robots such as the Air Penguin, Air

Ray, and Air Jelly have lighter-than-air bodies, propelled by paddles, and guided by sonar.

Snaking

Several snake robots have been successfully developed. Mimicking the way real snakes move, these robots can navigate very confined spaces, meaning they may one day be used to search for people trapped in collapsed buildings.The Japanese ACM-R5 snake robot can even navigate both on land and in water.

Skating

A small number of skating robots have been developed, one of which is a multi- mode walking and skating device. It has four legs, with unpowered wheels, which can either step or roll.Another robot, Plen, can use a miniature skateboard or roller-skates, and skate across a desktop.

Human-robot interaction

Kismet can produce a range of facial expressions. The state of the art in sensory intelligence for robots will have to progress through several orders of magnitude if we want the robots working in our homes to go beyond vacuum-cleaning the floors. If robots are to work effectively in homes and other non-industrial environments, the way they are instructed to perform their jobs, and especially how they will be told to stop will be of critical importance. The people who interact with them may have little or no training in robotics, and so any interface will need to be extremely intuitive. Science fiction authors also typically assume that robots will eventually be capable of communicating with humans through speech, gestures, and facial expressions, rather than a command-line interface. Although speech would be the most natural way for the human to communicate, it is unnatural for the robot. It will probably be a long time before robots interact as naturally as the fictional C-3PO, or Data of Star Trek, Next Generation.

Speech recognition

Interpreting the continuous flow of sounds coming from a human, in real time, is a difficult task for a computer, mostly because of the great variability of speech. The same word, spoken by the same person may sound different depending on local acoustics, volume, the previous word, whether or not the speaker has a cold, etc.. It becomes even harder when the speaker has a different accent. Nevertheless, great strides have been made in the field since Davis, Biddulph, and Balashek designed the first "voice input system" which recognized "ten digits spoken by a single user with 100% accuracy" in 1952.Currently, the best systems can recognize continuous, natural speech, up to 160 words per minute, with an accuracy of 95%.With the help of artificial intelligence, machines nowadays can use people's voice to identify their emotions such as satisfied or angry

Robotic voice

Other hurdles exist when allowing the robot to use voice for interacting with humans. For social reasons, synthetic voice proves suboptimal as a communication medium,making it necessary to develop the emotional component of robotic voice through various techniques. An advantage of diphonic branching is the emotion that the robot is programmed to project, can be carried on the voice tape, or phoneme, already pre-programmed onto the voice media. One of the earliest examples is a teaching robot named leachim developed in 1974 by Michael J. Freeman. Leachim was able to convert digital memory to rudimentary verbal speech on pre-recorded computer discs.It was programmed to teach students in The Bronx, New York.

Gestures

One can imagine, in the future, explaining to a robot chef how to make a pastry, or asking directions from a robot police officer. In both of these cases, making hand gestures would aid the verbal descriptions. In the first case, the robot would be recognizing gestures made by the human, and perhaps repeating them for confirmation. In the second case, the robot police officer would gesture to indicate "down the road, then turn right". It is likely that gestures will make up a part of the interaction between humans and robots.A great many systems have been developed to recognize human hand gestures.

Facial expression

Facial expressions can provide rapid feedback on the progress of a dialog between two humans, and soon may be able to do the same for humans and robots. Robotic faces have been constructed by Hanson Robotics using their elastic polymer called Frubber, allowing a large number of facial expressions due to the elasticity of the rubber facial coating and embedded subsurface motors (servos). The coating and servos are built on a metal skull. A robot should know how to approach a human, judging by their facial expression and body language. Whether the person is happy, frightened, or crazy-looking affects the type of interaction expected of the robot. Likewise, robots like Kismet and the more recent addition, Nexi can produce a range of facial expressions, allowing it to have meaningful social exchanges with humans.

Artificial emotions

Artificial emotions can also be generated, composed of a sequence of facial expressions and/or gestures. As can be seen from the movie Final Fantasy: The Spirits Within, the programming of these artificial emotions is complex and requires a large amount of human observation. To simplify this programming in the movie, presets were created together with a special software program. This decreased the amount of time needed to make the film. These presets could possibly be transferred for use in real-life robots.

Personality

Many of the robots of science fiction have a personality, something which may or may not be desirable in the commercial robots of the future.Nevertheless, researchers are trying to create robots which appear to have a personality:i.e. they use sounds, facial expressions, and body language to try to convey an internal state, which may be joy, sadness, or fear. One commercial example is Pleo, a toy robot dinosaur, which can exhibit several apparent emotions.

Chapter 16: Social Intelligence

The Socially Intelligent Machines Lab of the Georgia Institute of Technology researches new concepts of guided teaching interaction with robots. The aim of the projects is a social robot that learns task and goals from human demonstrations without prior knowledge of high-level concepts. These new concepts are grounded from low-level continuous sensor data through unsupervised learning, and task goals are subsequently learned using a Bayesian approach. These concepts can be used to transfer knowledge to future tasks, resulting in faster learning of those tasks. The results are demonstrated by the robot Curi who can scoop some pasta from a pot onto a plate and serve the sauce on top. Aircraft (Zhuhai) Co., Ltd. had successfully developed a bionic gecko robot named "Speedy Freelander". According to Dr. Yeung, the gecko robot could rapidly climb up and down a variety of building walls, navigate through ground and wall fissures, and walk upside-down on the ceiling. It was also able to adapt to the surfaces of smooth glass, rough, sticky or dusty walls as well as various types of metallic materials. It could also identify and circumvent obstacles automatically. Its flexibility and speed were comparable to a natural gecko. A third approach is to mimic the motion of a snake climbing a pole.

Autonomy levels

Control systems may also have varying levels of autonomy. Direct interaction is used for haptic or teleoperated devices, and the human has nearly complete control over the robot's motion. Operator-assist modes have the operator commanding medium-to-high-level tasks, with the robot automatically figuring out how to achieve them. An autonomous robot may go without human interaction for extended periods of time . Higher levels of autonomy do not necessarily require more complex cognitive capabilities. For example, robots in assembly plants are completely autonomous but operate in a fixed pattern. Another classification takes into account the interaction between human control and the machine motions.

Teleoperation. A human controls each movement, each machine actuator change is specified by the operator.

Supervisory. A human specifies general moves or position changes and the machine decides specific movements of its actuators.

Task-level autonomy. The operator specifies only the task and the robot manages itself to complete it.

Full autonomy. The machine will create and complete all its tasks without human interaction.

Research

Two Jet Propulsion Laboratory engineers stand with three vehicles, providing a size comparison of three generations of Mars rovers. Front and center is the flight spare for the first Mars rover, Sojourner, which landed on Mars in 1997 as part of the Mars Pathfinder Project. On the left is a Mars Exploration Rover (MER) test vehicle that is a working sibling to Spirit and Opportunity, which landed on Mars in 2004. On the right is a test rover for the Mars Science Laboratory, which landed Curiosity on Mars in 2012.Sojourner is 65 cm (2.13 ft) long. The Mars Exploration Rovers (MER) are 1.6 m (5.2 ft) long. Curiosity on the right is 3 m (9.8 ft) long. Much of the research in robotics focuses not on specific industrial tasks, but on investigations into new types of robots, alternative ways to think about or design robots, and new ways to manufacture them. Other investigations, such as MIT's cyberflora project, are almost wholly academic. A first particular new innovation in robot design is the open sourcing of robot-projects. To describe the level of advancement of a robot, the term "Generation Robots" can be used. This term is coined by Professor Hans Moravec,

Principal Research Scientist at the Carnegie Mellon University Robotics Institute in describing the near future evolution of robot technology. First generation robots, Moravec predicted in 1997, should have an intellectual capacity comparable to perhaps a lizard and should become available by 2010. Because the first generation robot would be incapable of learning, however, Moravec predicts that the second generation robot would be an improvement over the first and become available by 2020, with the intelligence maybe comparable to that of a mouse. The third generation robot should have the intelligence comparable to that of a monkey. Though fourth generation robots, robots with human intelligence, professor Moravec predicts, would become possible, he does not predict this happening before around 2040 or 2050. The second is evolutionary robots. This is a methodology that uses evolutionary computation to help design robots, especially the body form, or motion and behavior controllers. In a similar way to natural evolution, a large population of robots is allowed to compete in some way, or their ability to perform a task is measured using a fitness function. Those that perform worst are removed from the population and replaced by a new set, which have new behaviors based on those of the winners. Over time the population improves, and eventually a satisfactory robot may appear. This happens without any direct programming of the robots by the researchers. Researchers use this method both to create better robots, and to explore the nature of evolution. Because the process often requires many generations of robots to be simulated,this technique may be run entirely or mostly in simulation, using a robot simulator software package, then tested on real robots once the evolved algorithms are good enough.Currently, there are about 10 million industrial robots toiling around the world, and Japan is the top country having high density of utilizing robots in its manufacturing industry.

Dynamics and kinematics

The study of motion can be divided into kinematics and dynamics. Direct kinematics or forward kinematics refers to the calculation of end effector position, orientation, velocity, and acceleration when the corresponding joint values are known. Inverse kinematics refers to the opposite case in which required joint values are calculated for given end effector values, as done in path planning. Some special aspects of kinematics include handling of redundancy (different possibilities of performing the same movement), collision avoidance, and singularity avoidance. Once all relevant positions, velocities, and accelerations have been calculated using kinematics, methods from the field of dynamics are used to study the effect of forces upon these movements. Direct dynamics refers to the calculation of accelerations in the robot once the applied forces are known. Direct dynamics is used in computer simulations of the robot. Inverse dynamics refers to the calculation of the actuator forces necessary to create a prescribed end-effector acceleration. This information can be used to improve the control algorithms of a robot. In each area mentioned above, researchers strive to develop new concepts and strategies, improve existing ones, and improve the interaction between these areas. To do this, criteria for "optimal" performance and ways to optimize design, structure, and control of robots must be developed and implemented.

Bionics and biomimetics

Bionics and biomimetics apply the physiology and methods of locomotion of animals to the design of robots. For example, the design of BionicKangaroo was based on the way kangaroos jump

Education and training

Robotics engineers design robots, maintain them, develop new applications for them, and conduct research to expand the potential of robotics.Robots have become a popular educational tool in some middle and high schools, particularly in parts of the USA,as well as in numerous youth summer camps, raising interest in programming, artificial intelligence, and robotics among students.

Career training

Universities like Worcester Polytechnic Institute (WPI) offer bachelors, masters, and doctoral degrees in the field of robotics. Vocational schools offer robotics training aimed at careers in robotics.

Chapter 17: Certification

The Robotics Certification Standards Alliance (RCSA) is an international robotics certification authority that confers various industry- and educational-related robotics certifications.

Summer robotics camp

Several national summer camp programs include robotics as part of their core curriculum. In addition, youth summer robotics programs are frequently offered by celebrated museums and institutions.

Robotics competitions

There are many competitions around the globe. The SeaPerch curriculum is aimed as students of all ages. This is a short list of competition examples; for a more complete list see Robot competition.

Competitions for Younger Children

The FIRST organization offers the FIRST Lego League Jr. competitions for younger children. This competition's goal is to offer younger children an opportunity to start learning about science and technology. Children in this competition build Lego models and have the option of using the Lego WeDo robotics kit.

Competitions for Children Ages 9-14

One of the most important competitions is the FLL or FIRST Lego League. The idea of this specific competition is that kids start developing knowledge and getting into robotics while playing with Lego since they are nine years old. This competition is associated with National Instruments. Children use Lego Mindstorms to solve autonomous robotics challenges in this competition.

Competitions for Teenagers

The FIRST Tech Challenge is designed for intermediate students, as a transition from the FIRST Lego League to the FIRST Robotics Competition. The FIRST Robotics Competition focuses more on mechanical design, with a specific game being played each year. Robots are built specifically for that year's game. In match play, the robot moves autonomously during the first 15 seconds of the game (although certain years such as 2019's Deep Space change this rule), and is manually operated for the rest of the match.

Competitions for Older Students

The various RoboCup competitions include teams of teenagers and university students. These competitions focus on soccer competitions with different types of robots, dance competitions, and urban search and rescue competitions. All of the robots in these competitions must be autonomous. Some of these competitions focus on simulated robots.

AUVSI runs competitions for flying robots, robot boats, and underwater robots.

The Student AUV Competition Europe (SAUC-E) mainly attracts undergraduate and graduate student teams. As in the AUVSI competitions, the robots must be fully autonomous while they are participating in the competition.

The Microtransat Challenge is a competition to sail a boat across the Atlantic Ocean.

Competitions Open to Anyone RoboGames is open to anyone wishing to compete in their over 50 categories of robot competitions. Federation of International Robot-soccer Association holds the FIRA World Cup competitions. There are flying robot competitions, robot soccer competitions, and other challenges, including weightlifting barbells made from dowels and CDs.

Robotics afterschool programs

Many schools across the country are beginning to add robotics programs to their after school curriculum. Some major programs for afterschool robotics include FIRST Robotics Competition, Botball and B.E.S.T. Robotics.Robotics competitions often include aspects of business and marketing as well as engineering and design. The Lego company began a program for children to learn and get excited about robotics at a young age.

Decolonial Educational Robotics

Decolonial Educational Robotics is a branch of Decolonial Technology, and Decolonial A.I.,practiced in various places around the world. This methodology is summarized in pedagogical theories and practices such as Pedagogy of the Oppressed and Montessori methods. And it aims at teaching robotics from the local culture, to pluralize and mix technological knowledge

Employment

Robotics is an essential component in many modern manufacturing environments. As factories increase their use of robots, the number of robotics–related jobs grow and have been observed to be steadily rising.The employment of robots in industries has increased productivity and efficiency savings and is typically seen as a long-term investment for benefactors. A paper by Michael Osborne and Carl Benedikt Frey found that 47 per cent of US jobs are at risk to automation "over some unspecified number of years".These claims have been criticized on the ground that social policy, not AI, causes unemployment.In a 2016 article in The Guardian, Stephen Hawking stated "The automation of factories has already decimated jobs in traditional manufacturing, and the rise of artificial intelligence is likely to extend this job destruction deep into the middle classes, with only the most caring, creative or supervisory roles remaining".